Workbook

Progress in
Mathematics

SADLIER-OXFORD

Rose Anita McDonnell

Catherine D. LeTourneau

Anne Veronica Burrows

Francis H. Murphy

M. Winifred Kelly

with

Dr. Elinor R. Ford

Series Consultants

Tim Mason
Math Specialist
Palm Beach County School District
West Palm Beach, FL

Margaret Mary Bell, S.H.C.J., Ph.D
Director, Teacher Certification
Rosemont College
Rosemont, PA

Dennis W. Nelson, Ed.D.
Director of Basic Skills
Mesa Public Schools
Mesa, AZ

Sadlier-Oxford
A Division of William H. Sadlier, Inc.

Table of Contents

Textbook Chapter	Workbook Page

| Textbook Chapter | Workbook Page | Textbook Chapter | Workbook Page |

Problem-Solving Strategy: Choose the Operation

Name _____

Date _____

Caroline feeds her horse 2 quarts of grain a day. How many quarts of grain does he eat in 5 days?

Think: You are joining equal quantities. Multiply.

$5 \times 2 = 10$

The horse eats 10 quarts of grain.

Solve. Do your work on a separate sheet of paper.

1. There are 16 horses stabled at Shady Oaks Stable. The horses can go out to the pasture in groups of 8. How many groups do they make?

2. George works at the stable on weekends. He worked 4 hours on Saturday and 5 hours on Sunday. How many hours did he work that weekend?

3. Rob is training his horse for a special event. He rides for two hours a day, 6 days a week. How many hours does he ride during one week?

4. Amy cleaned two stalls on Monday. She cleaned three times as many on Tuesday. How many did she clean on Tuesday?

5. Gina exercises some of the horses. She rode 7 horses on Thursday and 5 horses on Friday. How many more horses did she ride on Thursday?

6. Lucky Lady spends 4 hours a day In the pasture. How many hours does she spend in the pasture during one week?

7. There are 9 brown horses and 6 white horses. How many more brown horses are there?

8. The vet gave shots to 5 horses on Monday and 6 horses on Tuesday. How many horses in all got shots?

 1

Problem-Solving Strategy: Guess and Test

Name _____

Date _____

There are 12 families at the hiking club picnic. Some families have 3 people. The rest have 5 people each. There are 44 people in all. How many families have 5 people? How many families have 3 people?

Guess the number of 5-person families: 5.   Think: $5 \times 5 = 25$.

Make a table to record your guesses.

5-People Families	3-People Families	Number of People	Test
$5 \times 5 = 25$	$7 \times 3 = 21$	$25 + 21 = 46$	too high
$3 \times 5 = 15$	$9 \times 3 = 27$	$15 + 27 = 42$	too low
$4 \times 5 = 20$	$8 \times 3 = 24$	$20 + 24 = 44$	correct

Solve. Do your work on a separate sheet of paper.

1. Bonnie has 5 more books than Timothy. Together they have 23 books. How many books does Timothy have? How many books does Bonnie have?

   _____

2. Shiying has twice as many dimes as nickels. Altogether he has $2.00. How many dimes does Shiying have? How many nickels does he have?

   _____

3. There are 29 members at the debate club meet. The judges divide all the members into 9 teams. There are 4 members on some teams, and 3 members on each of the rest. How many teams have 4 members? How many teams have 3 members?

   _____

4. Darlene spends 33 hours of her vacation hiking. She hikes every day of her 7-day vacation. Some days she hikes for 4 hours, and some she hikes for 5 hours. How many days does she hike for 4 hours? How many days does she hike for 5 hours?

   _____

5. Karen counts 58 people at the town meeting. There are 6 more women than men. How many men are there? How many women are there?

   _____

6. Walter works 10 fewer hours than Maggie. Together they work 34 hours. How many hours does Maggie work? How many hours does Walter work?

   _____

Problem-Solving Strategy: Information from a Table

Name _____

Date _____

The Kaplan family is vacationing in Michigan. They plan to drive from Kalamazoo to Ann Arbor. How many miles will they drive?

Mileage Beween Cities in Michigan

	Flint	Detroit	Ann Arbor	Lansing
Grand Rapids	107	156	131	66
Kalamazoo	126	142	97	75
Battle Creek	101	113	75	50
Ironwood	555	613	607	550
Jackson	88	80	35	38

The Kaplans will drive 97 miles from Kalamazoo to Ann Arbor.

Solve. Use the table above. Do your work on a separate sheet of paper.

1. On Friday morning, the Kaplans drove from Battle Creek to Lansing. Then they drove 127 more miles. How many miles did they drive on Friday?

2. Of all the cities listed in the chart, which two are the greatest distance from each other? How far apart are they?

3. The Kaplans have relatives in Ann Arbor. They are now in Ironwood. How far away are their relatives?

4. On Wednesday, the Kaplans drove from Battle Creek to Detroit. How far did they drive?

5. The Kaplans are in Jackson. They are trying to decide whether to drive to Flint or Lansing. How much farther is the trip to Flint?

6. One day the Kaplans drove from Detroit to Grand Rapids. How many miles did they drive?

7. Andrew took a bus from Jackson to Ann Arbor. How many miles was the bus ride?

8. The Kaplans are in Jackson. Are they closer to Flint or Detroit?

3

Problem-Solving Strategy: Write a Number Sentence

Name _____

Date _____

Kelly agreed to score each rider's test for 8 classes at the pony club horse show. There were 6 riders in each class. How many tests did she score?

Multiply to find the product. $8 \times 6 = 48$

Kelly scored 48 tests.

Think: Because there are 8 sets of 6, multiply.
$8 \times 6 =$?

Write a number sentence. Then solve.
Do your work on a separate sheet of paper.

1. Susanne has 35 horse magazines. William has 12 more horse magazines than Susanne. How many horse magazines does William have?

   _____

2. Juan bought a saddle pad for $23 and a book on horse grooming for $5. How much did he spend for both the saddle pad and the book?

   _____

3. Wei rides her horse 4 days each week. She rides for the same length of time each day. If she rides for 12 hours each week, how many hours does she ride each day?

   _____

4. Mr. Zhao is a horse photographer. At the spring horse show, he took four pictures of each of 8 riders. How many pictures did he take?

   _____

5. There are 6 turnout fields on the Winding Creek Horse Farm. There are 9 horses in each field. How many horses are there altogether?

   _____

6. Angelina's pony, Rusty, has a stall that is 8 feet long and 6 feet wide. What is the perimeter of the stall?

   _____

7. Denise bought a book on quarter horses for $6 and a book on thoroughbreds for $17. How much more did she pay for the book on thoroughbreds?

   _____

8. Tommy rides a horse twice a week. He rode for 50 minutes on Wednesday and 75 minutes on Saturday. How much longer did he ride on Saturday than on Wednesday?

   _____

Thousands

Name _____

Date _____

THOUSANDS PERIOD			ONES PERIOD		
h	t	o	h	t	o
4	3	4,	8	0	9

Standard Form: 434,809
Word Name: four hundred thirty-four thousand, eight hundred nine

What is the value of the underlined digit?

1. 2<u>5</u>2 _____

2. 81<u>2</u> _____

3. <u>4</u>11 _____

4. 6,<u>9</u>39 _____

5. 7,82<u>4</u> _____

6. 3,6<u>9</u>5 _____

7. <u>9</u>,326 _____

8. 4,<u>5</u>16 _____

9. <u>3</u>6,945 _____

10. 8<u>1</u>,469 _____

11. <u>3</u>19,675 _____

12. 6<u>5</u>4,389 _____

Write the letter to answer each question.

a. 168,003   **b.** 710,304   **c.** 780,452   **d.** 806,145

13. Which number shows 6 thousands? _____

14. Which number shows 8 ten thousands? _____

15. Which numbers have no tens? _____

Write the number in standard form.

16. forty thousand, eight hundred sixty-five _____

17. three hundred sixty-two thousand, one hundred seven _____

18. six hundred seventy-nine thousand _____

19. nine thousand _____

20. five hundred thousand, two _____

Write the number in words.

21. 7,014 _____

22. 400,684 _____

23. 569,081 _____

Millions

Name _____

Date _____

MILLIONS PERIOD	THOUSANDS PERIOD	ONES PERIOD

hundreds | tens | ones | hundreds | tens | ones | hundreds | tens | ones

| 2 | 6, | 7 | 0 | 9, | 3 | 8 | 0 |

Standard Form: 26,709,380

Word Name: twenty-six million, seven hundred nine thousand, three hundred eighty

Write each digit of the number in its correct place in the chart.

MILLIONS PERIOD			THOUSANDS PERIOD			ONES PERIOD		
h	t	o	h	t	o	h	t	o

1. 9,239,124

2. 47,962,471

3. 625,452,548

Write the period of the underlined digits.

4. <u>963</u>,479 _____

5. 836,<u>592</u> _____

6. <u>806</u>,219,479 _____

7. 259,<u>724</u>,416 _____

Write the value of the underlined digit.

8. <u>6</u>,479,219 _____

9. 35,07<u>4</u>,250 _____

10. 8<u>6</u>3,592 _____

11. <u>9</u>15,291,801 _____

Write in standard form.

12. seventy-six million, fifty-five thousand, two hundred eighty _____

13. five hundred eight million, two hundred seven thousand, nine _____

14. four million, three hundred thousand, four hundred twenty-five _____

15. fifteen million, six thousand, one hundred two _____

16. thirty-one million, seven hundred two _____

Place Value

Name _____

Date _____

Standard Form:	Expanded Form:
4,381,256	4,000,000 + 300,000 + 80,000 + 1000 + 200 + 50 + 6

Write each number in expanded form.

1. 79 _____ **2.** 4096 _____

3. 987 _____ **4.** 341 _____

5. 37,420 _____ **6.** 500,707 _____

7. 838,292 _____

8. 1,023,152 _____

9. 8,795,123 _____

10. 10,193,764 _____

11. 43,780,383 _____

12. 503,010,208 _____

Write each number in standard form.

13. 60 + 8 _____ **14.** 200 + 40 + 2 _____

15. 3000 + 300 + 30 + 3 _____ **16.** 70,000 + 2000 + 500 + 60 + 3 _____

17. 100,000 + 20,000 + 3000 + 500 + 70 + 9 _____

18. 700,000 + 10,000 + 4000 + 800 + 30 + 1 _____

19. 6,000,000 + 300,000 + 80,000 + 5000 + 100 + 70 + 3 _____

20. 40,000,000 + 6,000,000 + 800,000 + 100 + 1 _____

Write the numbers that are 1000 more and 1000 less.

21. 576,233 _____ **22.** 47,073,963 _____

23. 32,470 _____ **24.** 503,000,019 _____

25. 46,852 _____ **26.** 35,184,251 _____

Comparing and Ordering Whole Numbers

Name _____

Date _____

Compare: 7635 _?_ 7653	Order from least to greatest:
Compare from left to right until you find digits that are not the same.	2163, 2430, 2621, 2330
	The thousands are the same.
Compare the tens: $3 < 5$	Compare hundreds and rearrange.
So 7635 $<$ 7653.	2163, 2330, 2403, 2621

Compare. Write <, =, or >.

1. 539____935

2. 2690____2609

3. 8174____747

4. 2,627____2,726

5. 1,415____1,454

6. 35,452____35,524

7. 9584____5489

8. 9962____9862

9. 81,604____81,605

10. 8119____8109

11. 32,909____32,909

12. 16,421____16,420

13. 14,487____14,478

14. 780____8708

15. 12,321____13,221

16. 34,826____34,862

17. 5124____65,142

18. 17,532____12,357

19. 6215____6215

20. 17,519____51,917

21. 92,592____93,592

22. 1000____10,000

23. 6536____5263

24. 76,919____76,019

Write in order from least to greatest.

25. 48,964; 48,649; 46,894; 49,446 _____; _____; _____; _____

26. 2345; 2342; 2376; 2321; 2365 _____; _____; _____; _____; _____

Write in order from greatest to least.

27. 6501; 4565; 8537; 9570 _____; _____; _____; _____

28. 43,833; 43,733; 43,893; 43,863 _____; _____; _____; _____

PROBLEM SOLVING

29. Denver made 3951 field goals.
    Los Angeles made 3964 field goals.
    Which basketball team made more
    field goals?

    _____

Number Sense:
Using a Number Line

Name _____

Date _____

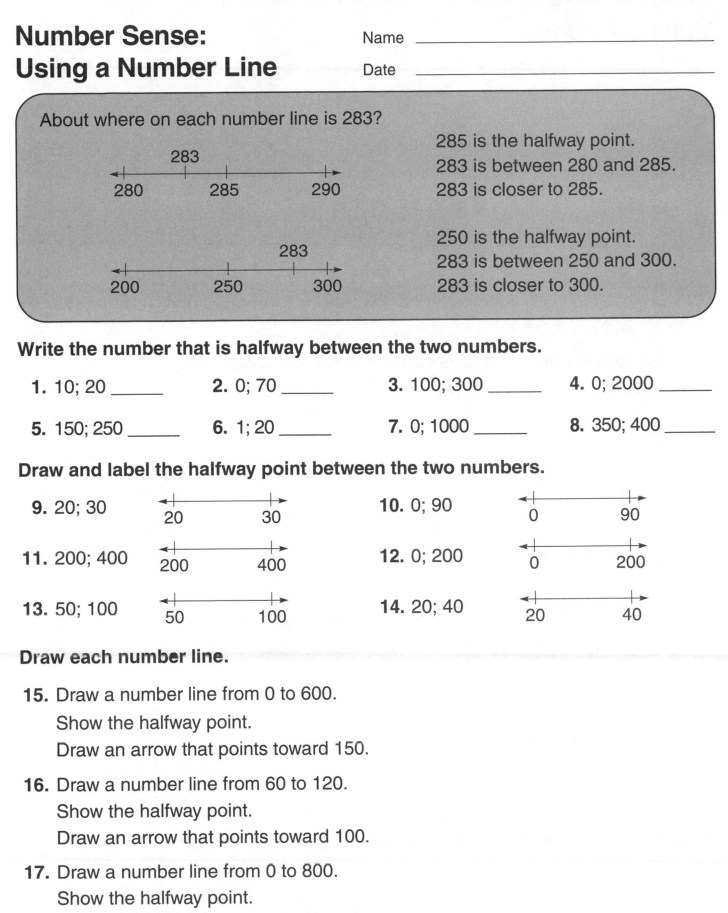

About where on each number line is 283?

283

280   285   290

285 is the halfway point.
283 is between 280 and 285.
283 is closer to 285.

283

200   250   300

250 is the halfway point.
283 is between 250 and 300.
283 is closer to 300.

Write the number that is halfway between the two numbers.

1. 10; 20 _____   **2.** 0; 70 _____   **3.** 100; 300 _____   **4.** 0; 2000 _____

5. 150; 250 _____   **6.** 1; 20 _____   **7.** 0; 1000 _____   **8.** 350; 400 _____

Draw and label the halfway point between the two numbers.

9. 20; 30   20   30

10. 0; 90   0   90

11. 200; 400   200   400

12. 0; 200   0   200

13. 50; 100   50   100

14. 20; 40   20   40

Draw each number line.

15. Draw a number line from 0 to 600.
Show the halfway point.
Draw an arrow that points toward 150.

16. Draw a number line from 60 to 120.
Show the halfway point.
Draw an arrow that points toward 100.

17. Draw a number line from 0 to 800.
Show the halfway point.
Draw an arrow that points toward 225.

Use with Lesson 1-7, text pages 48–49.   9

Making Change

Name _____

Date _____

Count up to make change.

Cost: Amount Given:

$7.69 ⟶ $7.70 ⟶ $7.75 ⟶ $8.00 ⟶ $9.00 ⟶ $10.00

Count to find the value of the change.

$1.00 + $1.00 + $.25 + $.05 + $.01

$1.00 ⟶ $2.00 ⟶ $2.25 ⟶ $2.30 ⟶ $2.31

Write the fewest coins and bills you would receive as change.
Then write the value of the change.

1. Cost: $.57   Coins and Bills: _____

   Amount Given: $1.00   Value: _____

2. Cost: $1.98   Coins and Bills: _____

   Amount Given: $5.00   Value: _____

3. Cost: $3.45   Coins and Bills: _____

   Amount Given: $5.00   Value: _____

4. Cost: $9.23   Coins and Bills: _____

   Amount Given: $20.00   Value: _____

5. Cost: $14.95   Coins and Bills: _____

   Amount Given: $20.00   Value: _____

6. Cost: $17.29   Coins and Bills: _____

   Amount Given: $20.00   Value: _____

10   **Use with Lesson 1-8, text pages 50–51.**

Comparing and Ordering Money

Compare: $71.38 _?_ $71.83 Compare ten dollars. $70 = $70 Compare dollars. $1 = $1 Compare dimes. $.30 < $.80 So $71.38 < $71.83.	**Order from least to greatest:** $7.43, $7.49, $6.43 Compare dollars. $6 < $7 and $7 = $7 So $6.43 is least. Compare dimes. $.40 = $.40 Compare pennies. $.03 < $.09 $6.43, $7.43, $7.49

Compare. Write <, =, or >.

1. $.54___$.45

2. $.18___$.81

3. $1.46___$1.34

4. $3.84___$4.83

5. $14.25___$14.25

6. $32.45___$32.25

7. $4.48___$4.85

8. $2.68___$2.95

9. $10.01___$11.00

10. $54.14___$52.14

11. $77.07___$77.70

12. $16.28___$15.82

13. $.66___$.55

14. $9.99___$10.99

15. $24.35___$12.36

16. $20.00___$30.00

17. $59.01___$59.01

18. $65.43___$76.54

Write in order from least to greatest.

19. $6.63, $6.53, $6.68, $6.56 _____, _____, _____, _____

20. $17.43, $17.34, $17.45, $17.41 _____, _____, _____, _____

Write in order from greatest to least.

21. $3.35, $3.37, $3.27, 3.75 _____, _____, _____, _____

22. $46.80, $36.80, $26.81, $58.60 _____, _____, _____, _____

Rounding

Round numbers to tell about how much or about how many.
Round to the nearest ten.

57 ⟶ 60   7 > 5 Round up.

Round to the nearest hundred.

1453 ⟶ 1500   5 = 5 Round up.

Round to the nearest thousand. 7489 ⟶ 7000   4 < 5 Round down.

Round to the nearest ten or ten cents.

1. 67 _____   **2.** 81 _____   **3.** 44 _____

4. 188 _____   **5.** 248 _____   **6.** 92 _____

7. $.87 _____   **8.** $.15 _____   **9.** $.52 _____

10. $5.28 _____   **11.** $9.45 _____   **12.** $8.52 _____

Round to the nearest hundred or dollar.

13. 678 _____   **14.** 871 _____   **15.** 250 _____

16. 1323 _____   **17.** 5348 _____   **18.** 4693 _____

19. $2.56 _____   **20.** $6.79 _____   **21.** $8.15 _____

22. $12.42 _____   **23.** $52.51 _____   **24.** $46.89 _____

Round to the nearest thousand or ten dollars.

25. 4799 _____   **26.** 6666 _____   **27.** 18,512 _____

28. $23.45 _____   **29.** $357.39 _____   **30.** $439.99 _____

Problem-Solving Strategy: Make a Table or List

Name _____

Date _____

Lucy planted 25 flowers. She planted
4 tulips for every iris.
How many tulips did she plant?

Make a table. Multiply each number of
irises by 4 to find the number of tulips.
Then add to find the number of flowers.

Type	Number of Flowers				
Iris	1	2	3	4	5
Tulip	4	8	12	16	20
Total	5	10	15	20	25

Lucy planted 20 tulips.

Solve. Do your work on a separate sheet of paper.

1. Liza is mixing blue paint with yellow paint to make green paint. She uses 1 quart of blue paint for every 5 quarts of yellow paint. How many quarts of yellow paint will she need to make 36 quarts of green paint?

2. Trenton has 2 hats and 6 T-shirts. His hats are blue and red. His T-shirts are white, green, black, yellow, purple, and orange. How many ways can he wear the hats and shirts together?

3. Abby, Ben and Carol are having their picture taken. In how many different ways can they arrange themselves in a line from left to right for the photo?

4. Tina and Bruce are each rolling a 1 to 6 number cube. They are looking for different ways to roll two factors whose product is greater than 14. How many different ways will they find?

5. Elizabeth has 45¢ in her pocket. None of the coins are pennies. Make a list of all the combinations that could be in Elizabeth's pocket.

6. Derek is ordering lunch. He can choose either a tuna, an egg, or a chicken sandwich. He can also choose either vegetable, pea, or chicken noodle soup. How many different ways can Derek order a sandwich and soup for lunch?

Addition Properties

Name _____

Date _____

Change the order.

$$
\begin{array}{r}
2 \\
2 \\
1 \\
+\,9 \\
\hline
14
\end{array}
\quad
\begin{array}{l}
14 \\
12 \\
10
\end{array}
$$

Change the order and the grouping.

$$
\begin{array}{r}
2 \\
3 \\
+\,7 \\
\hline
12
\end{array}
$$

$(3+7) + 2 = 12$

$10 \quad + 2 = 12$

Add. Use the addition properties.

1. $\begin{array}{r}0 \\ +\,7 \\ \hline\end{array}$	**2.** $\begin{array}{r}7 \\ +\,0 \\ \hline\end{array}$	**3.** $\begin{array}{r}8 \\ +\,9 \\ \hline\end{array}$	**4.** $\begin{array}{r}9 \\ +\,8 \\ \hline\end{array}$	**5.** $\begin{array}{r}6 \\ +\,5 \\ \hline\end{array}$	**6.** $\begin{array}{r}5 \\ +\,6 \\ \hline\end{array}$
7. $\begin{array}{r}4 \\ +\,7 \\ \hline\end{array}$	**8.** $\begin{array}{r}7 \\ +\,4 \\ \hline\end{array}$	**9.** $\begin{array}{r}5 \\ +\,8 \\ \hline\end{array}$	**10.** $\begin{array}{r}4 \\ +\,0 \\ \hline\end{array}$	**11.** $\begin{array}{r}7 \\ +\,5 \\ \hline\end{array}$	**12.** $\begin{array}{r}8 \\ +\,6 \\ \hline\end{array}$
13. $\begin{array}{r}4 \\ 5 \\ 3 \\ +\,7 \\ \hline\end{array}$	**14.** $\begin{array}{r}2 \\ 5 \\ 0 \\ +\,5 \\ \hline\end{array}$	**15.** $\begin{array}{r}2 \\ 9 \\ 1 \\ +\,3 \\ \hline\end{array}$	**16.** $\begin{array}{r}4 \\ 2 \\ 4 \\ +\,2 \\ \hline\end{array}$	**17.** $\begin{array}{r}1 \\ 6 \\ 4 \\ +\,4 \\ \hline\end{array}$	**18.** $\begin{array}{r}2 \\ 0 \\ 6 \\ +\,1 \\ \hline\end{array}$

Add the number in the center to each number around it.
Write each addition sentence.

19.

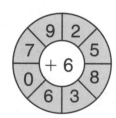

20.

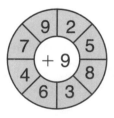

Use with Lesson 2-1, text pages 68–69.

Addition Strategies

Name _____

Date _____

Use doubles.	Use 10.
5 + 6 = ?	9 + 6 = ?
5 + 5 = 10	10 + 6 = 16
5 + 6 = 11	9 + 6 = 15

$$\left.\begin{array}{r} 4 \\ 3 \\ 1 \\ +3 \\ \hline 11 \end{array}\right\} 6 + 4 + 1 = 11$$

$$\left.\begin{array}{r} 3 \\ 1 \\ 7 \\ +2 \\ \hline 13 \end{array}\right\} 10 + 3 = 13$$

Find the sum.

1.	2.	3.	4.	5.	6.
4 + 5	9 + 8	6 + 5	7 + 7	7 + 8	4 + 9

Add mentally.

7.	8.	9.	10.	11.	12.
2 8 + 5	4 1 + 9	1 6 + 9	3 3 + 4	5 2 + 5	2 4 + 8

13.	14.	15.	16.	17.	18.
2 0 8 + 5	5 5 5 + 5	7 1 3 + 1	4 4 2 + 5	3 4 2 + 1	1 0 9 + 9

PROBLEM SOLVING

19. Luke found 6 scallop shells, 7 oyster shells and 4 clam shells on the beach. How many shells did Luke find altogether? _____

20. Tawana made a sculpture with 7 large shells and 6 small shells. How many shells did she use in all? _____

Subtraction Concepts

Take away.

5 − 2 = 3

Part of a whole set.

12 − 3 = 9

Compare.

4 − 3 = 1

How many more are needed?

8 − 4 = 4

Solve. Tell which meaning of subtraction is shown.

1. Team A scored 14 runs and Team B scored 6 runs. How many more runs did Team A score?

   _____

2. Cy had 12 boxes of popcorn and sold 5 of them. How many boxes were left?

   _____

3. Elena has 6 quarters. She needs 13 quarters to buy a T-shirt. How many more quarters does she need?

   _____

4. In the basketball game, Jamal scored 12 points and Luis scored 7 points. How many more points did Jamal score than Luis?

   _____

5. Sam had 10 action figures. Peter had 5 action figures. How many more did Sam have than Peter?

   _____

6. Tamika had 11 letters. She had enough stamps to mail 5 letters. How many letters were *not* mailed?

   _____

7. Tai wants to collect 12 baseball pennants. She has 3 pennants. How many more does she need?

   _____

8. Karen won a total of 10 ribbons at horse shows. Two ribbons were blue ribbons. How many were *not* blue ribbons?

   _____

Addition and Subtraction Sentences

Name _____

Date _____

Find the missing addend: $7 + \underline{?} = 13$	Find the missing minuend: $9 = \underline{?} - 4$
$13 - 7 = \underline{?}$	$9 + 4 = \underline{?}$
Think: $13 - 7 = 6$	Think: $9 + 4 = 13$
So $7 + 6 = 13$.	So $9 = 13 - 4$.

Find the missing addend.

1. $6 + \underline{\hphantom{00}} = 10$ **2.** $\underline{\hphantom{00}} + 9 = 12$ **3.** $9 + \underline{\hphantom{00}} = 18$

4. $7 + \underline{\hphantom{00}} = 14$ **5.** $10 + \underline{\hphantom{00}} = 17$ **6.** $8 + \underline{\hphantom{00}} = 13$

7. $4 + \underline{\hphantom{00}} = 12$ **8.** $4 + \underline{\hphantom{00}} = 11$ **9.** $7 + \underline{\hphantom{00}} = 15$

Find the minuend or subtrahend.

10. $6 = \underline{\hphantom{00}} - 9$ **11.** $14 - \underline{\hphantom{00}} = 5$ **12.** $15 - \underline{\hphantom{00}} = 9$

13. $7 = \underline{\hphantom{00}} - 7$ **14.** $8 = 13 - \underline{\hphantom{00}}$ **15.** $9 - \underline{\hphantom{00}} = 0$

16. $5 = \underline{\hphantom{00}} - 5$ **17.** $12 - \underline{\hphantom{00}} = 8$ **18.** $17 - \underline{\hphantom{00}} = 8$

Find the missing number.

19.
$$\begin{array}{r} 4 \\ + \\ \hline 11 \end{array}$$
20.
$$\begin{array}{r} \\ - 7 \\ \hline 8 \end{array}$$
21.
$$\begin{array}{r} 11 \\ - \\ \hline 9 \end{array}$$
22.
$$\begin{array}{r} 9 \\ + \\ \hline 14 \end{array}$$
23.
$$\begin{array}{r} \\ + 6 \\ \hline 12 \end{array}$$
24.
$$\begin{array}{r} 17 \\ - \\ \hline 8 \end{array}$$

25.
$$\begin{array}{r} 13 \\ - \\ \hline 9 \end{array}$$
26.
$$\begin{array}{r} \\ - 8 \\ \hline 8 \end{array}$$
27.
$$\begin{array}{r} \\ + 7 \\ \hline 16 \end{array}$$
28.
$$\begin{array}{r} 6 \\ + \\ \hline 11 \end{array}$$
29.
$$\begin{array}{r} 5 \\ + \\ \hline 13 \end{array}$$
30.
$$\begin{array}{r} \\ - 7 \\ \hline 5 \end{array}$$

PROBLEM SOLVING

31. Don had a roll of film with 12 pictures. He took 4 pictures. How many pictures were left on the roll?

_____

Mental Math

Think of tens or hundreds.

2400	24 hundreds
− 600	− 6 hundreds
1800	18 hundreds

Look for a pattern.

155	155	155
+ 10	+ 20	+ 30
165	175	185

Look for pairs of numbers that add to 10 or 100.

Add: $42 + 20 + 80$

$$\begin{array}{r} 42 \\ 20 \\ + 80 \\ \hline 142 \end{array}\Big\rangle 100 + 42 = 142$$

Add mentally.

1. $30 + 90 =$ _____

2. $40 + 40 =$ _____

3. $70 + 60 =$ _____

4. $600 + 400 =$ _____

5. $1200 + 700 =$ _____

6. $3500 + 300 =$ _____

Subtract mentally.

7. $90 - 10 =$ _____

8. $70 - 50 =$ _____

9. $800 - 700 =$ _____

10. $560 - 60 =$ _____

11. $6800 - 600 =$ _____

12. $8500 - 500 =$ _____

Add or subtract mentally.

13.
$$\begin{array}{r} 719 \\ + 10 \\ \hline \end{array}$$

14.
$$\begin{array}{r} 719 \\ + 20 \\ \hline \end{array}$$

15.
$$\begin{array}{r} 719 \\ + 30 \\ \hline \end{array}$$

16.
$$\begin{array}{r} 719 \\ + 40 \\ \hline \end{array}$$

17.
$$\begin{array}{r} 719 \\ + 50 \\ \hline \end{array}$$

18.
$$\begin{array}{r} 719 \\ + 60 \\ \hline \end{array}$$

19.
$$\begin{array}{r} 600 \\ - 50 \\ \hline \end{array}$$

20.
$$\begin{array}{r} 500 \\ - 50 \\ \hline \end{array}$$

21.
$$\begin{array}{r} 400 \\ - 50 \\ \hline \end{array}$$

22.
$$\begin{array}{r} 300 \\ - 50 \\ \hline \end{array}$$

23.
$$\begin{array}{r} 200 \\ - 50 \\ \hline \end{array}$$

24.
$$\begin{array}{r} 100 \\ - 50 \\ \hline \end{array}$$

25.
$$\begin{array}{r} 3 \\ 6 \\ + 7 \\ \hline \end{array}$$

26.
$$\begin{array}{r} 5 \\ 4 \\ + 5 \\ \hline \end{array}$$

27.
$$\begin{array}{r} 2 \\ 8 \\ + 7 \\ \hline \end{array}$$

28.
$$\begin{array}{r} 10 \\ 90 \\ + 41 \\ \hline \end{array}$$

29.
$$\begin{array}{r} 70 \\ 77 \\ + 30 \\ \hline \end{array}$$

30.
$$\begin{array}{r} 19 \\ 40 \\ + 60 \\ \hline \end{array}$$

Use with Lesson 2-5, text pages 76–77.

Estimating Sums and Differences

Name _____

Date _____

Use rounding to estimate sums and differences.

Estimate: 3271 + 190 + 6044
Round to hundreds.

$$3217 \longrightarrow 3200$$
$$190 \longrightarrow 200$$
$$+ 6044 \longrightarrow + 6000$$
$$\text{about} \quad 9400$$

Estimate: $27.99 − $3.14
Round to dollars.

$$\$27.99 \longrightarrow \$28.00$$
$$- \quad 3.14 \longrightarrow - \quad 3.00$$
$$\text{about} \quad \$25.00$$

Estimate the sum.

1. $9.19
  + 5.62

2. $ .13
  + .75

3. 205
  +381

4. 38
  +32

5. $62.41
  + 6.66

6. 3118
  2846
  +1979

7. $5.22
  .55
  + .13

8. 905
  19
  +727

9. $44.11
  2.25
  + 11.34

10. $10.45
  34.50
  + 3.17

Estimate the difference.

11. 92
  −49

12. $13.90
  − 7.32

13. $8.48
  − 2.68

14. 619
  −491

15. $.87
  − .54

Is the answer reasonable? Estimate to check. Then write *Yes* or *No*.

16. 42 + 27 = 69 _____

17. $69.08 − $41.45 = $27.63 _____

18. 5219 − 1523 = 4996 _____

19. 72 − 31 = 21 _____

20. $27.94 + $9.95 = $47.89 _____

21. 3402 − 306 = 3096 _____

22. $11.25 + $2.96 + $42.05 = $62.26 _____

23. 1412 + 2392 + 3114 = 5818 _____

Adding and Subtracting Money

Name _____

Date _____

Add.	Estimate.		Subtract.	Estimate.
$72.02 →	$70.00		$58.45 →	$60.00
+ 16.95 →	+ 20.00		− 27.04 →	− 30.00
$88.97	about $90.00		$31.41	about $30.00

Estimate. Then add.

1. $.31
+ .24

2. $.66
+ .11

3. $.44
+ .12

4. $7.63
+ 2.35

5. $1.25
+ 6.30

6. $4.30
+ 5.65

7. $4.32
+ 3.47

8. $16.90
+ 73.04

9. $47.52
+ 42.27

10. $28.30
+ 51.50

11. $5.21 + $2.03 + $1.14 = _____   **12.** $.20 + $.46 + $.32 = _____

Estimate. Then subtract.

13. $.89
− .09

14. $.97
− .16

15. $7.85
− 2.40

16. $6.97
− 3.15

17. $9.49
− 5.35

18. $3.79
− 1.67

19. $8.96
− 7.23

20. $17.85
− 10.23

21. $97.66
− 45.41

22. $28.75
− 26.75

23. $9.97 − $.91 = _____   **24.** $37.98 − $16.76 = _____

PROBLEM SOLVING

25. Beau spent $22.45 for a new pair of sneakers. Art spent $39.95 for a pair of boots. How much more did Art spend than Beau?

_____

Checking Addition and Subtraction

Name _____

Date _____

Add up:	Subtract one addend from the sum.	Add the difference and the subtrahend.

Add up:

→ 9986

   1032 ↑
   6511
+ 2443
→ 9986

Subtract one addend from the sum.

  342 ←   449
+ 107   − 107
  449 → 342

Add the difference and the subtrahend.

  786 ←   572
− 214   + 214
  572 → 786

Add and check.

1.   1364
      201
+ 6223

2. $10.15
   24.01
+ 62.72

3.   411
    250
+ 1327

4. $45.42
    2.23
+ 2.14

5.   425
+ 331   − 331
         425

6.   233
+ 614   − 614
         233

7.   710
+ 89   − 89
         710

Subtract and check.

8.   798
− 234   + 234
         798

9.   865
− 601   + 601
         865

10. $2.79
    .46 + .46
        $2.79

11.   488
− 365   + 365
         488

12. 987
− 624 + 624
        987

13. $6.75
− 6.23 + 6.23
      $6.75

Problem-Solving Strategy: Logical Reasoning

Name _____

Date _____

> Moe has a green lunchbox. He buys milk at school. Sue's lunchbox matches her thermos. Ron buys lunch and brings juice in a thermos. Whose thermos is the red one filled with milk?
>
> Sue has the red thermos.
>
> **Think:** Moe does not use a thermos.
> Ron's thermos has juice.
> Sue must have the red thermos.

Solve. Do your work on a separate sheet of paper.

1. Desmond gave Brenda 7 coins worth 92¢. Two coins were quarters. What were the other 5 coins?

2. Dawn, Doug, and Dale earned different amounts: $9.35, $8.52, and $8.25. Dale and Doug each earned about $9.00. Dawn earned exactly $1.10 less than Dale. How much did each earn?

3. Ian has 6 nature books. For each nature book, he has 2 science books. For every science book, he has 2 animal books. How many animal books does he have?

4. Chandra, Jane, Jim, and Alec are standing in line. Jim is behind Chandra and is not in front of Alec. Chandra is not first. Jane is last. In what order are the children standing in line?

5. Nick wears jersey number 42. Al's jersey has an even number that is greater than Nick's. Joe's number is greater than Al's and is odd. Stan's number, 46, is less than Joe's. If all of the numbers are between 41 and 48, what is each boy's jersey number?

6. A new jacket costs $50. A customer pays with a combination of 10-, 5-, and 1-dollar bills. She pays with 5 of one kind of bill, and the same number of each of the other bills. What combination of bills does she give the cashier?

Front-End Estimation

Name _____

Date _____

Estimate:	Then adjust:	Estimate:
3726	3**726** + 9**344** + 1089	$73.29
9344	⎴ ⎴	−44.36
+ 1089	about 1000	about $30.00
about **13**,000	13,000 + 1000 = 14,000	

Make a rough estimate on the first line. Then adjust on the second line.

1.	283	2.	498	3.	738	4.	$3.25	5.	$4.33
	409		126		344		7.67		6.49
	+ 135		+ 219		+ 106		+ 3.21		+ 2.97
	____		____		____		____		____
	____		____		____		____		____

6.	1356	7.	3195	8.	9238	9.	$56.74	10.	$66.39
	4160		1645		1721		43.23		53.26
	+ 4522		+ 3872		+ 4230		+ 39.72		+ 35.12
	____		____		____		____		____
	____		____		____		____		____

11. 482 + 247 + 166 + 135 _____ _____

12. $3.39 + $6.67 + $2.87 + $4.45 _____ _____

Estimate the difference. Use front-end estimation.

13.	441	14.	$9.37	15.	3642	16.	$82.02	17.	$55.66
	− 163		− 2.69		− 1829		− 16.55		− 40.78

PROBLEM SOLVING Use the table.

18. Ann organized a tag sale in her neighborhood. Two friends brought things to sell. About how much money did the three friends make?

_____

Tag Sale Income	
Ann	$16.61
Rhoda	$21.87
Bruce	$14.45

Adding with Regrouping

Name _____

Date _____

$147 + 282 = \underline{\ ?\ }$

Add ones, then tens, then hundreds.
Regroup as necessary.

h	t	o
^{1}1	4	7
+ 2	8	2
4	2	9

Estimate. Then add.

1. 62
    + 28

2. 49
    + 9

3. 51
    + 6

4. 77
    + 14

5. 38
    + 37

6. 714
    + 239

7. 681
    + 149

8. 236
    + 673

9. 168
    + 596

10. 328
    + 193

11. $5.28
    + 3.57

12. $3.98
    + 4.32

13. 325
    + 467

14. 325
    + 594

15. $1.38
    + 6.75

16. $518 + 293 = $ _____

17. $645 + 288 = $ _____

18. $296 + 377 = $_____

19. $\$4.83 + \$1.58 = $ _____

20. $348 + 57 = $ _____

21. $719 + 198 = $ _____

PROBLEM SOLVING

22. Before noon, 153 people visited the Central Post Office. After noon, 288 people visited. How many people visited the Post Office that day?

    _____

23. Edward sold 198 rolls of stamps in the morning and 79 rolls in the afternoon. How many rolls did he sell in all?

    _____

Three- and Four-Digit Addition

Name _____

Date _____

$6{,}443 + 9{,}489 = \underline{\ ?\ }$

Add the ones. Regroup.	Add the tens. Regroup.	Add the hundreds.	Add the thousands.

th	h	t	o
		1	
6	4	4	3
+9	4	8	9
			2

th	h	t	o
	1	1	
6	4	4	3
+9	4	8	9
		3	2

th	h	t	o
	1	1	
6	4	4	3
+9	4	8	9
	9	3	2

th	h	t	o
	1	1	
6	4	4	3
+9	4	8	9
15	9	3	2

Estimate. Then add.

1.	7543 + 1384	2.	7024 + 653	3.	6753 + 5908	4.	5724 + 5543
5.	741 + 625	**6.**	456 + 857	**7.**	453 + 647	**8.**	775 + 860
9.	7426 + 3021	**10.**	586 + 520	**11.**	9163 + 4248	**12.**	707 + 449
13.	281 + 945	**14.**	8876 + 5239	**15.**	$73.52 + 50.14	**16.**	$5.12 + 8.56

Align and add.

17. 776 + 668 = _____

18. 416 + 720 = _____

19. $58.07 + $56.94 = _____

20. $6.92 + $5.43 = _____

21. 8301 + 7289 = _____

22. 9533 + 6418 = _____

PROBLEM SOLVING

23. At the book auction, the Center City Public Library bought 5478 books. The University Library bought 6357 books. How many books did both libraries buy?

_____

Three or More Addends

Name _____

Date _____

$2247 + 316 + 539 = \underline{\ ?\ }$

Add the ones. Regroup.	Add the tens. Regroup.	Add the hundreds.	Add the thousands.

th	h	t	o
		2	
2	2	4	7
	3	1	6
+	5	3	9
			2

th	h	t	o
	1	2	
2	2	4	7
	3	1	6
+	5	3	9
		0	2

th	h	t	o
	1	2	
2	2	4	7
	3	1	6
+	5	3	9
	1	0	2

th	h	t	o
1	1	2	
2	2	4	7
	3	1	6
+	5	3	9
3	1	0	2

Estimate. Then find the sum.

1. $\begin{array}{r} 70 \\ 24 \\ 15 \\ + 32 \\ \hline \end{array}$

2. $\begin{array}{r} 46 \\ 83 \\ 10 \\ + 52 \\ \hline \end{array}$

3. $\begin{array}{r} 16 \\ 9 \\ 34 \\ + 21 \\ \hline \end{array}$

4. $\begin{array}{r} \$ \ .21 \\ .17 \\ .33 \\ + \ .56 \\ \hline \end{array}$

5. $\begin{array}{r} \$1.73 \\ .04 \\ .97 \\ + \ .25 \\ \hline \end{array}$

6. $\begin{array}{r} 753 \\ 942 \\ 214 \\ + 357 \\ \hline \end{array}$

7. $\begin{array}{r} 127 \\ 435 \\ 815 \\ + 393 \\ \hline \end{array}$

8. $\begin{array}{r} 234 \\ 678 \\ 89 \\ + \ \ 5 \\ \hline \end{array}$

9. $\begin{array}{r} \$3.77 \\ 6.01 \\ .74 \\ + \ .08 \\ \hline \end{array}$

10. $\begin{array}{r} \$6.92 \\ 3.25 \\ .96 \\ + 8.02 \\ \hline \end{array}$

11. $\begin{array}{r} 2343 \\ 3416 \\ + 2355 \\ \hline \end{array}$

12. $\begin{array}{r} 4129 \\ 2083 \\ + 9611 \\ \hline \end{array}$

13. $\begin{array}{r} 3214 \\ 2106 \\ + 6072 \\ \hline \end{array}$

14. $\begin{array}{r} \$24.94 \\ 18.06 \\ + 62.13 \\ \hline \end{array}$

15. $\begin{array}{r} \$13.52 \\ 84.40 \\ + 25.66 \\ \hline \end{array}$

Align and add.

16. $5763 + 4117 + 51 = $ _____

17. $\$25.24 + \$3.18 + \$.79 + \$1.04 = $ _____

PROBLEM SOLVING

18. Maria bought a pair of socks for $2.79, a shirt for $15.98 and a skirt for $24.95. How much did she spend in all?

_____

Subtracting with Regrouping

Name _____

Date _____

$671 - 352 = \underline{\ ?\ }$

Regroup

Subtract the ones.	Subtract the tens.	Subtract the hundreds.
h t o	h t o	h t o
6̸ 7̸ⁱ¹ 1	6̸ 7̸ⁱ¹ 1	6̸ 7̸ⁱ¹ 1
6 7 1	6 7 1	6 7 1
− 3 5 2	− 3 5 2	− 3 5 2
9	1 9	3 1 9

Check by adding. $352 + 319 = 671$

Estimate. Then find the difference. Check by adding.

1.	56 − 17	**2.**	90 − 48	**3.**	47 − 29	**4.**	$.81 − .56	**5.**	$.63 − .35

6.	$2.58 − 1.39	**7.**	$4.41 − 4.08	**8.**	996 − 589	**9.**	813 − 431	**10.**	624 − 252

11.	877 − 296	**12.**	684 − 49	**13.**	$.91 − .06	**14.**	$4.47 − .86

15. 85 − 8

Align and subtract.

16. $95 - 8 = $ _____

17. $63 - 9 = $ _____

18. $380 - 62 = $ _____

19. $591 - 88 = $ _____

20. $\$8.49 - \$.95 = $ _____

21. $\$9.35 - \$.73 = $ _____

PROBLEM SOLVING

22. The Busy Bake Shop had 192 loaves of bread. They sold 87 loaves in the morning. How many were left?

_____

Subtraction: Regrouping Twice

Name _____

Date _____

$844 - 297 = \underline{?}$

More ones needed.
Regroup.
Subtract ones.

$$\begin{array}{r} 8\,4\,\overset{3\ 14}{\cancel{4}} \\ -\,2\,9\,7 \\ \hline 7 \end{array}$$

More tens needed.
Regroup.
Subtract tens.

$$\begin{array}{r} 8\,\overset{7\ \overset{13}{\cancel{3}}\ 14}{4}\,\cancel{4} \\ -\,2\,9\,7 \\ \hline 4\,7 \end{array}$$

Subtract
hundreds.

$$\begin{array}{r} \overset{7\ \overset{13}{\cancel{3}}\ 14}{8}\,4\,\cancel{4} \\ -\,2\,9\,7 \\ \hline 5\,4\,7 \end{array}$$

Check by adding. $547 + 297 = 844$

Estimate. Then subtract. Check by adding.

1.	2.	3.	4.	5.
634 − 487	736 − 258	573 − 396	745 − 498	825 − 176

6.	7.	8.	9.	10.
326 − 268	413 − 247	764 − 389	843 − 754	921 − 237

11.	12.	13.	14.	15.
$3.52 − 1.93	$5.42 − .47	$4.63 − 1.95	$9.52 − 2.78	$7.27 − 1.59

Align and subtract.

16. $742 - 378 = $ _____

17. $6.20 - $3.35 = $ _____

18. $623 - 58 = $ _____

19. $563 - 396 = $ _____

20. $8.13 - $.26 = $ _____

21. $418 - 259 = $ _____

PROBLEM SOLVING

22. Dan collected 435 stamps. He gave away 187 of them. How many stamps did he have left?

_____

Three- and Four-Digit Subtraction

Name _____

Date _____

$2531 - 1665 = \underline{\ ?\ }$

Subtract. Regroup as needed.

$$
\begin{array}{r}
{\scriptstyle 2\ 11} \\
2\,5\,3\,\cancel{1} \\
-\,1\,6\,6\,5 \\
\hline
8\,6\,6
\end{array}
\qquad
\begin{array}{r}
{\scriptstyle 12} \\
{\scriptstyle 4\ \cancel{2}\ 11} \\
2\,5\,3\,\cancel{1} \\
-\,1\,6\,6\,5 \\
\hline
8\,6\,6
\end{array}
\qquad
\begin{array}{r}
{\scriptstyle 14\ 12} \\
{\scriptstyle 1\ \cancel{4}\ \cancel{2}\ 11} \\
2\,5\,3\,\cancel{1} \\
-\,1\,6\,6\,5 \\
\hline
8\,6\,6
\end{array}
\qquad
\begin{array}{r}
{\scriptstyle 14\ 12} \\
{\scriptstyle 1\ \cancel{4}\ \cancel{2}\ 11} \\
2\,5\,3\,\cancel{1} \\
-\,1\,6\,6\,5 \\
\hline
8\,6\,6
\end{array}
$$

Check by adding. $866 + 1665 = 2531$

Estimate. Then subtract. Check by adding.

1. $\begin{array}{r} 873 \\ -\ 594 \\ \hline \end{array}$	**2.** $\begin{array}{r} 556 \\ -\ 378 \\ \hline \end{array}$	**3.** $\begin{array}{r} 627 \\ -\ 149 \\ \hline \end{array}$	**4.** $\begin{array}{r} \$8.65 \\ -\ 2.98 \\ \hline \end{array}$	**5.** $\begin{array}{r} \$3.74 \\ -\ 1.96 \\ \hline \end{array}$
6. $\begin{array}{r} 9522 \\ -\ 3569 \\ \hline \end{array}$	**7.** $\begin{array}{r} 4226 \\ -\ 1332 \\ \hline \end{array}$	**8.** $\begin{array}{r} \$15.13 \\ -\ 6.95 \\ \hline \end{array}$	**9.** $\begin{array}{r} \$54.91 \\ -\ 17.76 \\ \hline \end{array}$	**10.** $\begin{array}{r} \$72.65 \\ -\ 25.88 \\ \hline \end{array}$

Align and subtract.

11. $321 - 124 =$ _____

12. $\$6.12 - \$3.79 =$ _____

13. $\$44.15 - \$15.44 =$ _____

14. $\$89.23 - \$29.75 =$ _____

15. $8776 - 2980 =$ _____

16. $9251 - 5423 =$ _____

17. $485 - 69 =$ _____

18. $4792 - 2843 =$ _____

PROBLEM SOLVING

19. It costs $45.95 to ride the bus from Albion to Los Rios. From Los Rios to Palmer, the bus costs $62.65. How much more does the bus to Palmer cost?

_____

Zeros in Subtraction

Name _____

Date _____

$800 - 425 =$ _?_

When there are zeros in the minuend, you may need to regroup more than once before you start to subtract.

```
    7 10              7 9 10            7 9 10
    8̶ 0̶ 0            8̶ 0̶ 0̶            8̶ 0̶ 0̶
  − 4 2 5          − 4 2 5          − 4 2 5
                                      3 7 5
```

Check by adding: $375 + 425 = 800$

Regroup. Then find the difference. Check by adding.

1. 3 0 0
    − 1 1 4

2. 8 0 0
    − 6 5 4

3. $2.0 0
    − 1.1 9

4. $7.0 0
    − 1.2 9

5. 4 0 3
    − 2 3 8

6. 7 0 5
    − 3 6 8

7. $2.0 4
    − 1.8 9

8. $6.0 8
    − 3.2 9

9. 3 0 3 0
    − 1 9 6 7

10. 6 0 0 0
    − 4 0 7 5

11. 7 0 0 0
    −   4 8 1

12. 6 0 0 4
    − 2 5 1 9

13. $9 0.0 0
    − 1 3.5 4

14. $5 0.0 5
    − 1 1.7 8

15. $1 8.0 9
    −   9.9 5

Align and subtract.

16. $600 - 142 =$ _____

17. $\$5.01 - \$2.97 =$ _____

18. $5000 - 974 =$ _____

19. $\$60.08 - \$9.62 =$ _____

20. $1300 - 516 =$ _____

21. $7043 - 5126 =$ _____

PROBLEM SOLVING

22. Etta needs to have $30.00 to buy a computer game. She has already earned $17.25 washing cars. How much more money does she need to earn? _____

Larger Sums and Differences

Name _____

Date _____

Add or subtract from right to left. Regroup as necessary.

Check subtraction by adding.

Add: 14,287 + 22,905 = __?__

$$
\begin{array}{r}
^{1}^{1} \\
1\,4{,}2\,8\,7 \\
+\,2\,2{,}9\,0\,5 \\
\hline
3\,7{,}1\,9\,2
\end{array}
$$

Subtract: \$263.71 − \$25.99 = __?__

$$
\begin{array}{rr}
\$2\,6\,3.7\,1 & \$2\,3\,7.7\,2 \\
-2\,5.9\,9 & +2\,5.9\,9 \\
\hline
\$2\,3\,7.7\,2 & \$2\,6\,3.7\,1
\end{array}
$$

Add or subtract. Watch for + or −.

1. 1 7,8 5 8 + 1 0,2 4 0	2. 6 9,7 6 6 − 2 4,8 7 3	3. \$5 9 7.9 6 − 4 5.1 8	4. 3 5,4 2 9 + 1 6,9 0 7
5. 7 2,1 1 1 − 8,4 2 6	6. \$4 1 8.2 2 − 1 1 9.5 5	7. \$2 7 6.0 5 + 1 3 5.1 7	8. 6 3,2 4 0 − 4 8,5 1 7
9. 6 1,8 4 6 + 4 0,2 3 7	10. 3 8,5 1 1 − 2 5,7 3 5	11. \$7 2 7.0 4 + 1 6 4.5 8	12. \$9 1 6.4 0 − 2 4 1.6 8
13. 1 6,2 8 1 7 2,7 2 4 + 5 0,4 1 6	14. \$1 5 0.9 5 3 7.6 6 + 5 0 4.1 4	15. \$4 2 9.1 7 3.0 6 + 4 1.8 2	16. 3 2,6 7 9 8,4 1 2 + 8 1,3 0 5
17. 6 2,0 0 4 − 3 5,2 5 8	18. 5 0,0 0 0 − 1 1,8 6 2	19. \$7 0 0.0 0 + 9 2.1 8	20. \$8 0 0.4 6 − 2 5.7 5

Align. Add or subtract.

21. 49,254 + 3,121 + 6,048 = _____

22. \$400.00 − \$253.61 = _____

23. \$723.29 + \$7.11 + \$.84 = _____

24. 87,400 − 2,946 = _____

Problem-Solving Strategy: Extra Information

Name _____

Date _____

The museum had 2419 visitors in the month of December.
The average donation was $3.00 per visitor. There were 450 fewer
visitors in January than in December. How many people visited
the museum in January?

What information is unnecessary?

The average donation was $3.00.

$2419 - 450 = 1969$

In January, 1969 people
visited the museum.

Solve. Do your work on a separate sheet of paper.

1. A crowd of 6240 saw the new show at the Metropolitan Art Gallery. In June 611 people came, and 1173 people came in July. How many more people came to the show in July?

2. In the Science Museum Gift Shop, Leslie bought a prism for $5.95. Walter bought a magnifier for $4.29 and two post cards for $.25 each. How much did Walter spend?

3. Lily and Luke went to the Children's Museum and built sculptures with blocks. Lily used 546 red blocks and 377 yellow blocks. Luke used 752 red blocks and 119 yellow blocks. How many red blocks did they use in all?

4. Skip's class went to the Natural History Museum and looked at some baskets that were made about 3000 years ago. Sahlah bought two other baskets that each cost $2.95. How much did she spend?

5. Malik's ticket to the Museum cost $3.50. He arrived at 12:00 noon and gave the clerk $5.00. How much change did he receive?

6. Charles bought a dinosaur puzzle with 675 pieces. Judy bought a puzzle with 950 pieces that cost $6.95. How many more pieces did Judy's puzzle have?

Multiplication Properties

Name _____

Date _____

Order Property	Property of One	Zero Property

$$\begin{array}{r} 2 \\ \times\,3 \\ \hline 6 \end{array} \qquad \begin{array}{r} 3 \\ \times\,2 \\ \hline 6 \end{array} \qquad\qquad \begin{array}{r} 1 \\ \times\,4 \\ \hline 4 \end{array} \qquad \begin{array}{r} 4 \\ \times\,1 \\ \hline 4 \end{array} \qquad\qquad \begin{array}{r} 0 \\ \times\,6 \\ \hline 0 \end{array} \qquad \begin{array}{r} 6 \\ \times\,0 \\ \hline 0 \end{array}$$

Complete.

1. $9 \times 8 = \underline{\hspace{1cm}} \times 9$

2. $\underline{\hspace{1cm}} \times 7 = 7 \times 4$

3. $6 \times 5 = 5 \times \underline{\hspace{1cm}}$

4. $5 \times 9 = \underline{\hspace{1cm}}$
$9 \times 5 = \underline{\hspace{1cm}}$

5. $6 \times 4 = \underline{\hspace{1cm}}$
$4 \times 6 = \underline{\hspace{1cm}}$

6. $7 \times 1 = \underline{\hspace{1cm}}$
$1 \times 7 = \underline{\hspace{1cm}}$

Find the product.

7. $\begin{array}{r} 6 \\ \times\,2 \\ \hline \end{array}$
8. $\begin{array}{r} 0 \\ \times\,5 \\ \hline \end{array}$
9. $\begin{array}{r} 3 \\ \times\,5 \\ \hline \end{array}$
10. $\begin{array}{r} 8 \\ \times\,0 \\ \hline \end{array}$
11. $\begin{array}{r} 9 \\ \times\,1 \\ \hline \end{array}$
12. $\begin{array}{r} 9 \\ \times\,8 \\ \hline \end{array}$

13. $\begin{array}{r} 7 \\ \times\,3 \\ \hline \end{array}$
14. $\begin{array}{r} 2 \\ \times\,1 \\ \hline \end{array}$
15. $\begin{array}{r} 7 \\ \times\,8 \\ \hline \end{array}$
16. $\begin{array}{r} 5 \\ \times\,6 \\ \hline \end{array}$
17. $\begin{array}{r} 7 \\ \times\,0 \\ \hline \end{array}$
18. $\begin{array}{r} 6 \\ \times\,9 \\ \hline \end{array}$

19. $1 \times 8 = \underline{\hspace{1cm}}$

20. $8 \times 1 = \underline{\hspace{1cm}}$

21. $9 \times 0 = \underline{\hspace{1cm}}$

22. $0 \times 4 =$

23. $4 \times 9 = \underline{\hspace{1cm}}$

24. $9 \times 4 = \underline{\hspace{1cm}}$

Complete. Identify the property.

25. $4 \times (2 \times 1) = (\underline{\hspace{1cm}} \times 2) \times 1$

$4 \times \underline{\hspace{1cm}} = \underline{\hspace{1cm}} \times 1$

$\underline{\hspace{1cm}} = \underline{\hspace{1cm}}$ _____ property

26. $5 \times (3 + 2) = (5 \times \underline{\hspace{1cm}}) + (\underline{\hspace{1cm}} \times 2)$

$5 \times \underline{\hspace{1cm}} = \underline{\hspace{1cm}} + \underline{\hspace{1cm}}$

$\underline{\hspace{1cm}} = \underline{\hspace{1cm}}$ _____ property

Use with Lesson 4-1, text pages 126–127. 33

Missing Factors

Name _____

Date _____

$6 \times \underline{\ ?\ } = 48$ **Think:** 6 times what number is 48?

$6 \times 6 = 36$ too small
$6 \times 7 = 42$ too small
$6 \times 8 = 48$ just right!   So $6 \times 8 = 48$.

Complete.

1. $6 \times \underline{\hspace{1cm}} = 48$

2. $8 \times \underline{\hspace{1cm}} = 32$

3. $\underline{\hspace{1cm}} \times 9 = 81$

4. $\underline{\hspace{1cm}} \times 7 = 35$

5. $\underline{\hspace{1cm}} \times 5 = 30$

6. $7 \times \underline{\hspace{1cm}} = 63$

7. $\underline{\hspace{1cm}} \times 8 = 64$

8. $5 \times \underline{\hspace{1cm}} = 40$

9. $8 \times \underline{\hspace{1cm}} = 72$

10. $5 \times \underline{\hspace{1cm}} = 25$

11. $\underline{\hspace{1cm}} \times 7 = 56$

12. $9 \times \underline{\hspace{1cm}} = 36$

13. $49 = 7 \times \underline{\hspace{1cm}}$

14. $42 = \underline{\hspace{1cm}} \times 6$

15. $27 = \underline{\hspace{1cm}} \times 3$

16. $45 = 9 \times \underline{\hspace{1cm}}$

17. $28 = \underline{\hspace{1cm}} \times 4$

18. $54 = 6 \times \underline{\hspace{1cm}}$

19. $36 = \underline{\hspace{1cm}} \times 6$

20. $24 = 3 \times \underline{\hspace{1cm}}$

21. $24 = \underline{\hspace{1cm}} \times 6$

PROBLEM SOLVING

22. Mrs. Bradley needs 40 juice boxes. Juice boxes are sold in eight-packs. How many eight-packs should she buy?

_____

23. Eight bottles fit in a case. There are 48 bottles. How many cases are there?

_____

24. Abby had 3 boxes of cards. Each box had the same number of cards. She had 15 cards in all. How many cards were in each box?

_____

Use with Lesson 4-2, text pages 128–129.

Multiplication Models

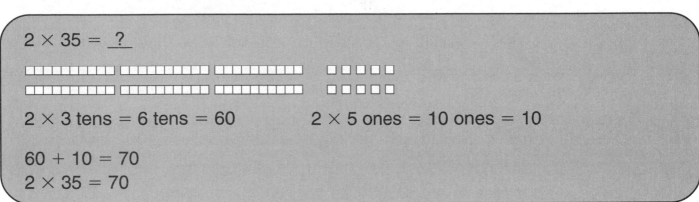

$2 \times 35 = \underline{\ ?\ }$

2×3 tens $= 6$ tens $= 60$ 2×5 ones $= 10$ ones $= 10$

$60 + 10 = 70$

$2 \times 35 = 70$

Write a multiplication sentence for each model.

1. _____

2. _____

3. _____

4. _____

PROBLEM SOLVING Use a model.

5. Pat makes quilts. She sewed 4 rows, each with 15 squares. How many squares did she use?

_____

6. Ms. Quintaro put 18 books on each of 5 new shelves in the library. How many books did she put away?

_____

7. Martin makes 3 trains with cubes. Each train uses 35 cubes. How many cubes does he use?

_____

8. Alison makes designs with paper shapes. She makes 6 rows with 10 shapes in each row. How many shapes does she use?

_____

9. Dwayne put 12 stamps on each of 8 pages in his book. How many stamps did he have?

_____

10. Raoul planted 6 rows of tomatoes. Each row had 18 plants. How many plants did he have?

_____

Special Factors

Look for patterns when you multiply tens, hundreds, or thousands.

$$40 \times 1 = 40 \quad \text{1 zero}$$

$$600 \times 3 = 1800 \quad \text{2 zeros}$$

$$4000 \times 6 = 24{,}000 \quad \text{3 zeros}$$

Complete.

1. 7×4 tens $= 7 \times 40 =$ _____

2. 3×9 tens $= 3 \times 90 =$ _____

3. 4×3 hundreds $= 4 \times$ _____ $=$ _____

4. 6×1 hundred $= 6 \times 100 =$ _____

5. 8×1 thousand $= 8 \times$ _____ $=$ _____

6. 5×2 thousands $=$ _____ $\times$ _____ $=$ _____

Multiply.

7.
$$\begin{array}{r} 80 \\ \times\ 3 \\ \hline \end{array}$$

8.
$$\begin{array}{r} 40 \\ \times\ 2 \\ \hline \end{array}$$

9.
$$\begin{array}{r} 20 \\ \times\ 7 \\ \hline \end{array}$$

10.
$$\begin{array}{r} 50 \\ \times\ 9 \\ \hline \end{array}$$

11.
$$\begin{array}{r} 300 \\ \times\ 4 \\ \hline \end{array}$$

12.
$$\begin{array}{r} 100 \\ \times\ 5 \\ \hline \end{array}$$

13.
$$\begin{array}{r} 700 \\ \times\ 8 \\ \hline \end{array}$$

14.
$$\begin{array}{r} 200 \\ \times\ 6 \\ \hline \end{array}$$

15.
$$\begin{array}{r} 8000 \\ \times\ 6 \\ \hline \end{array}$$

16.
$$\begin{array}{r} 5000 \\ \times\ 3 \\ \hline \end{array}$$

17.
$$\begin{array}{r} 2000 \\ \times\ 9 \\ \hline \end{array}$$

18.
$$\begin{array}{r} 4000 \\ \times\ 7 \\ \hline \end{array}$$

PROBLEM SOLVING

19. Sandy ordered 7 boxes of cones for the Divinely Delicious Yogurt Shop. There are 100 cones in a box. How many cones did she order?

_____

20. Jack ordered 2 cartons of napkins. There are 1000 napkins in a carton. How many napkins did he order?

_____

36   **Use with Lesson 4-4, text pages 132–133.**

Multiplying by One-Digit Numbers

Name _____

Date _____

$3 \times 32 = \underline{\ ?\ }$

Multiply the ones.

```
  3 2
×   3
─────
    6
```

$3 \times 2 \text{ ones} = 6 \text{ ones}$

Multiply the tens.

```
  3 2
×   3
─────
  9 6
```

$3 \times 3 \text{ tens} = 9 \text{ tens}$

Multiply.

1. 24
  × 2

2. 14
  × 2

3. 41
  × 2

4. 12
  × 4

5. 44
  × 2

6. 34
  × 2

7. 11
  × 5

8. 54
  × 1

9. 13
  × 2

10. 31
   × 3

11. 21
   × 4

12. 13
   × 3

13. 12
   × 3

14. 31
   × 2

15. 33
   × 3

16. $2 \times 33 =$ _____

17. $3 \times 21 =$ _____

18. $4 \times 12 =$ _____

19. $2 \times 12 =$ _____

20. $4 \times 22 =$ _____

21. $9 \times 11 =$ _____

PROBLEM SOLVING

22. Daniella sold boxes of cards to earn money to go to camp. She sold 12 boxes each week for 3 weeks. How many boxes of cards did she sell? _____

23. Alonzo sold 14 boxes of cards each week for 2 weeks. How many boxes of cards did he sell? _____

24. There are 22 cards in each box. How many cards are in 3 boxes? _____

Products:
Front-End Estimation

Name _____

Date _____

$5 \times \$7.90 = \underline{?}$

Multiply the front digit of the greater factor.

$$\begin{array}{r} \$\boxed{7}.90 \\ \times \quad 5 \\ \hline \boxed{35} \end{array}$$

Write $ and . in the product.

Write 0s for the other digits.

$$\begin{array}{r} \$ \ 7.9\,0 \\ \times \quad 5 \\ \hline \$3\,5.0\,\boxed{0} \end{array}$$

Estimate the product.

1. $\begin{array}{r} 96 \\ \times\ 4 \\ \hline \end{array}$ **2.** $\begin{array}{r} 67 \\ \times\ 5 \\ \hline \end{array}$ **3.** $\begin{array}{r} 23 \\ \times\ 6 \\ \hline \end{array}$ **4.** $\begin{array}{r} 82 \\ \times\ 3 \\ \hline \end{array}$ **5.** $\begin{array}{r} 14 \\ \times\ 9 \\ \hline \end{array}$

6. $\begin{array}{r} 823 \\ \times\ 3 \\ \hline \end{array}$ **7.** $\begin{array}{r} 431 \\ \times\ 5 \\ \hline \end{array}$ **8.** $\begin{array}{r} 593 \\ \times\ 2 \\ \hline \end{array}$ **9.** $\begin{array}{r} 601 \\ \times\ 7 \\ \hline \end{array}$ **10.** $\begin{array}{r} 296 \\ \times\ 8 \\ \hline \end{array}$

11. $\begin{array}{r} 2395 \\ \times\ 4 \\ \hline \end{array}$ **12.** $\begin{array}{r} 1473 \\ \times\ 6 \\ \hline \end{array}$ **13.** $\begin{array}{r} 3857 \\ \times\ 2 \\ \hline \end{array}$ **14.** $\begin{array}{r} 8732 \\ \times\ 5 \\ \hline \end{array}$ **15.** $\begin{array}{r} 7016 \\ \times\ 7 \\ \hline \end{array}$

16. $\begin{array}{r} \$.86 \\ \times\ 3 \\ \hline \end{array}$ **17.** $\begin{array}{r} \$6.24 \\ \times\ 3 \\ \hline \end{array}$ **18.** $\begin{array}{r} \$5.95 \\ \times\ 6 \\ \hline \end{array}$ **19.** $\begin{array}{r} \$86.49 \\ \times\ 2 \\ \hline \end{array}$ **20.** $\begin{array}{r} \$72.99 \\ \times\ 5 \\ \hline \end{array}$

PROBLEM SOLVING Use estimation.

21. Will 3 hats cost more or less than $10.00?

_____

22. About how much will 2 pairs of socks cost?

_____

23. Will 2 hats cost more or less than 4 T-shirts?

_____

Crazy Days Sale!!	
Hats	$2.95
Socks	$3.25
T-Shirts	$2.09

Use with Lesson 4-6, text pages 136–137.

Multiplying with Regrouping

Name _____

Date _____

$4 \times 37 = \underline{?}$

Multiply the ones.
Regroup.

$$\begin{array}{r} \overset{2}{3}7 \\ \times\ 4 \\ \hline 8 \end{array}$$

Multiply the tens.
Add the regrouped tens.

$$\begin{array}{r} \overset{2}{3}7 \\ \times\ 4 \\ \hline 148 \end{array}$$

Estimate. Then multiply.

1. $\begin{array}{r} 38 \\ \times\ 2 \\ \hline \end{array}$
2. $\begin{array}{r} 15 \\ \times\ 6 \\ \hline \end{array}$
3. $\begin{array}{r} 27 \\ \times\ 3 \\ \hline \end{array}$
4. $\begin{array}{r} 59 \\ \times\ 2 \\ \hline \end{array}$
5. $\begin{array}{r} 67 \\ \times\ 3 \\ \hline \end{array}$
6. $\begin{array}{r} 96 \\ \times\ 3 \\ \hline \end{array}$

7. $\begin{array}{r} 35 \\ \times\ 8 \\ \hline \end{array}$
8. $\begin{array}{r} 98 \\ \times\ 7 \\ \hline \end{array}$
9. $\begin{array}{r} 89 \\ \times\ 6 \\ \hline \end{array}$
10. $\begin{array}{r} 46 \\ \times\ 9 \\ \hline \end{array}$
11. $\begin{array}{r} 27 \\ \times\ 5 \\ \hline \end{array}$
12. $\begin{array}{r} 96 \\ \times\ 4 \\ \hline \end{array}$

13. $\begin{array}{r} 49 \\ \times\ 6 \\ \hline \end{array}$
14. $\begin{array}{r} 98 \\ \times\ 6 \\ \hline \end{array}$
15. $\begin{array}{r} 78 \\ \times\ 3 \\ \hline \end{array}$
16. $\begin{array}{r} 47 \\ \times\ 9 \\ \hline \end{array}$
17. $\begin{array}{r} 67 \\ \times\ 4 \\ \hline \end{array}$
18. $\begin{array}{r} 43 \\ \times\ 5 \\ \hline \end{array}$

19. $\begin{array}{r} 77 \\ \times\ 6 \\ \hline \end{array}$
20. $\begin{array}{r} 48 \\ \times\ 9 \\ \hline \end{array}$
21. $\begin{array}{r} 57 \\ \times\ 7 \\ \hline \end{array}$
22. $\begin{array}{r} 72 \\ \times\ 5 \\ \hline \end{array}$
23. $\begin{array}{r} 83 \\ \times\ 9 \\ \hline \end{array}$
24. $\begin{array}{r} 94 \\ \times\ 8 \\ \hline \end{array}$

25. $5 \times 86 =$ _____
26. $8 \times 48 =$ _____
27. $6 \times 57 =$ _____

28. $4 \times 73 =$ _____
29. $8 \times 82 =$ _____
30. $7 \times 93 =$ _____

PROBLEM SOLVING

31. On Monday, 75 children attended each performance of a puppet show. There were 3 performances. How many children attended the show in all? _____

32. Mr. Felkamp planted 8 rows of tomato plants. There were 96 plants in a row. How many tomato plants were there in all? _____

Multipying Three-Digit Numbers

Name _____

Date _____

$3 \times 249 =$?

Multiply the ones. Regroup.	Multiply the tens. Add and regroup.	Multiply the hundreds. Add the regrouped hundred.
$\begin{array}{r} \overset{2}{2\,4\,9} \\ \times\ \ \ 3 \\ \hline 7 \end{array}$	$\begin{array}{r} \overset{1}{2}\,\overset{2}{4}\,9 \\ \times\ \ \ 3 \\ \hline 4\,7 \end{array}$	$\begin{array}{r} \overset{1}{2}\,\overset{2}{4}\,9 \\ \times\ \ \ 3 \\ \hline 7\,4\,7 \end{array}$

Estimate. Then multiply.

1. $\begin{array}{r} 4\,5\,7 \\ \times\ \ \ 2 \\ \hline \end{array}$
2. $\begin{array}{r} 3\,5\,3 \\ \times\ \ \ 2 \\ \hline \end{array}$
3. $\begin{array}{r} 2\,8\,3 \\ \times\ \ \ 3 \\ \hline \end{array}$
4. $\begin{array}{r} 1\,9\,8 \\ \times\ \ \ 4 \\ \hline \end{array}$
5. $\begin{array}{r} 1\,3\,7 \\ \times\ \ \ 5 \\ \hline \end{array}$

6. $\begin{array}{r} 8\,4\,3 \\ \times\ \ \ 3 \\ \hline \end{array}$
7. $\begin{array}{r} 6\,1\,2 \\ \times\ \ \ 7 \\ \hline \end{array}$
8. $\begin{array}{r} 2\,8\,5 \\ \times\ \ \ 6 \\ \hline \end{array}$
9. $\begin{array}{r} 7\,3\,6 \\ \times\ \ \ 4 \\ \hline \end{array}$
10. $\begin{array}{r} 8\,2\,3 \\ \times\ \ \ 8 \\ \hline \end{array}$

11. $\begin{array}{r} 8\,3\,8 \\ \times\ \ \ 6 \\ \hline \end{array}$
12. $\begin{array}{r} 4\,3\,8 \\ \times\ \ \ 4 \\ \hline \end{array}$
13. $\begin{array}{r} 5\,2\,8 \\ \times\ \ \ 7 \\ \hline \end{array}$
14. $\begin{array}{r} 6\,4\,3 \\ \times\ \ \ 9 \\ \hline \end{array}$
15. $\begin{array}{r} 8\,4\,2 \\ \times\ \ \ 6 \\ \hline \end{array}$

Find the product.

16. $4 \times 238 =$ _____
17. $9 \times 812 =$ _____
18. $2 \times 287 =$ _____

19. $7 \times 126 =$ _____
20. $5 \times 534 =$ _____
21. $6 \times 445 =$ _____

PROBLEM SOLVING

22. There are 365 days in a year. How many days are there in 2 years?

_____

23. Each office has 132 square feet of floor. How many square feet of floor are in 5 offices?

_____

24. The factors are 183 and 3. What is the product?

_____

Use with Lesson 4-8, text pages 140–141.

Multiplying Money

Name _____

Date _____

> $7 \times \$4.29 = \underline{\,?\,}$
>
> Multiply the same way you multiply whole numbers.
> Write a decimal point in the product two places from
> the right. Write the dollar sign.
>
> $$\begin{array}{r} \$\ 4.29 \\ \times\qquad 7 \\ \hline \$30.03 \end{array}$$

Estimate. Then multiply.

1. $\begin{array}{r} \$.52 \\ \times\quad 7 \\ \hline \end{array}$

2. $\begin{array}{r} \$.39 \\ \times\quad 5 \\ \hline \end{array}$

3. $\begin{array}{r} \$.92 \\ \times\quad 9 \\ \hline \end{array}$

4. $\begin{array}{r} \$.81 \\ \times\quad 6 \\ \hline \end{array}$

5. $\begin{array}{r} \$.67 \\ \times\quad 4 \\ \hline \end{array}$

6. $\begin{array}{r} \$4.41 \\ \times\quad 5 \\ \hline \end{array}$

7. $\begin{array}{r} \$5.95 \\ \times\quad 8 \\ \hline \end{array}$

8. $\begin{array}{r} \$3.07 \\ \times\quad 6 \\ \hline \end{array}$

9. $\begin{array}{r} \$6.34 \\ \times\quad 4 \\ \hline \end{array}$

10. $\begin{array}{r} \$2.25 \\ \times\quad 7 \\ \hline \end{array}$

11. $\begin{array}{r} \$8.50 \\ \times\quad 3 \\ \hline \end{array}$

12. $\begin{array}{r} \$6.81 \\ \times\quad 9 \\ \hline \end{array}$

13. $\begin{array}{r} \$9.15 \\ \times\quad 5 \\ \hline \end{array}$

14. $\begin{array}{r} \$7.30 \\ \times\quad 8 \\ \hline \end{array}$

15. $\begin{array}{r} \$3.39 \\ \times\quad 9 \\ \hline \end{array}$

16. $4 \times \$.69 =$ _____

17. $3 \times \$.85 =$ _____

18. $5 \times \$3.95 =$ _____

19. $2 \times \$8.29 =$ _____

20. $6 \times \$7.89 =$ _____

21. $9 \times \$1.65 =$ _____

PROBLEM SOLVING

22. Terri bought 5 paperback books.
    What was the total cost?

    _____

23. Hassan bought 6 magazines.
    What was the total cost?

    _____

24. Lupe bought 2 hardcover books.
    What was the total cost?

    _____

25. Patrick bought 4 magazines.
    What was the total cost?

    _____

Book Sale!	
Hardcovers	$6.95
Paperbacks	$2.59
Magazines	$1.25

Multiplying Four-Digit Numbers

Name _____

Date _____

$3 \times 5487 = \underline{?}$

First estimate.

$$\begin{array}{r} 5487 \\ \times \quad 3 \\ \hline \text{about } 15{,}000 \end{array}$$

Then multiply.

$$\begin{array}{r} \overset{1\ 2\ 2}{5487} \\ \times \quad 3 \\ \hline 16{,}461 \end{array}$$

Estimate. Then multiply.

1.
$$\begin{array}{r} 1463 \\ \times \quad 7 \\ \hline \end{array}$$

2.
$$\begin{array}{r} 4925 \\ \times \quad 2 \\ \hline \end{array}$$

3.
$$\begin{array}{r} 6893 \\ \times \quad 8 \\ \hline \end{array}$$

4.
$$\begin{array}{r} 2995 \\ \times \quad 6 \\ \hline \end{array}$$

5.
$$\begin{array}{r} 7549 \\ \times \quad 7 \\ \hline \end{array}$$

6.
$$\begin{array}{r} 5386 \\ \times \quad 6 \\ \hline \end{array}$$

7.
$$\begin{array}{r} 6287 \\ \times \quad 3 \\ \hline \end{array}$$

8.
$$\begin{array}{r} 9956 \\ \times \quad 3 \\ \hline \end{array}$$

9.
$$\begin{array}{r} 4527 \\ \times \quad 9 \\ \hline \end{array}$$

10.
$$\begin{array}{r} 3245 \\ \times \quad 5 \\ \hline \end{array}$$

11.
$$\begin{array}{r} \$51.15 \\ \times \quad 5 \\ \hline \end{array}$$

12.
$$\begin{array}{r} \$74.87 \\ \times \quad 4 \\ \hline \end{array}$$

13.
$$\begin{array}{r} \$24.93 \\ \times \quad 7 \\ \hline \end{array}$$

14.
$$\begin{array}{r} \$68.65 \\ \times \quad 8 \\ \hline \end{array}$$

15.
$$\begin{array}{r} \$53.72 \\ \times \quad 9 \\ \hline \end{array}$$

Find the product.

16. $2 \times 9487 =$ _____

17. $6 \times 8312 =$ _____

18. $4 \times 6125 =$ _____

19. $8 \times \$88.31 =$ _____

20. $9 \times \$47.32 =$ _____

21. $5 \times \$27.43 =$ _____

PROBLEM SOLVING

22. If 4 quarters equal one dollar, how many quarters equal 4726 dollars?

_____

23. A postal worker sorted 8475 letters each day. How many letters did she sort in 5 days?

_____

24. Mr. Harris bought 2 snow tires that cost $59.95 each. What was the total cost?

_____

Use with Lesson 4-10, text pages 144–145.

Patterns in Multiplication

Name _____

Date _____

Multiply.

1.	2.	3.	4.	5.
81 $\times\ 10$	42 $\times\ 10$	75 $\times\ 10$	70 $\times\ 10$	90 $\times\ 10$

6.	7.	8.	9.	10.
776 $\times\ \ 10$	535 $\times\ \ 10$	948 $\times\ \ 10$	260 $\times\ \ 10$	400 $\times\ \ 10$

11.	12.	13.	14.	15.
70 $\times\ 50$	80 $\times\ 40$	50 $\times\ 60$	67 $\times\ 80$	45 $\times\ 40$

16.	17.	18.	19.	20.
74 $\times\ 50$	28 $\times\ 70$	59 $\times\ 30$	33 $\times\ 70$	35 $\times\ 90$

21.	22.	23.	24.	25.
560 $\times\ \ 70$	600 $\times\ \ 80$	150 $\times\ \ 60$	300 $\times\ \ 90$	430 $\times\ \ 50$

Complete each pattern.

26. $1 \times 58 =$ _____

$10 \times 58 =$ _____

$10 \times 580 =$ _____

27. $9 \times 70 =$ _____

$90 \times 70 =$ _____

$90 \times 700 =$ _____

Estimating Products

Name _____

Date _____

To estimate products:
Round each factor to its greatest place. Then multiply.

$8.67 → $9.00	26 → 30	343 → 300
× 42 → × 40	× 34 → × 30	× 66 → × 70
about $360.00	about 900	about 21,000

Estimate the product.

1. 88	**2.** 74	**3.** 41	**4.** 85	**5.** 31
× 55	× 69	× 47	× 42	× 36

6. $.72	**7.** $.57	**8.** $.66	**9.** $.82	**10.** $.93
× 91	× 75	× 24	× 64	× 24

11. 647	**12.** 529	**13.** 295	**14.** 879	**15.** 922
× 38	× 76	× 88	× 56	× 44

16. $6.76	**17.** $3.25	**18.** $3.21	**19.** $5.67	**20.** $8.43
× 34	× 16	× 75	× 63	× 52

21. 46 × 537 _____   **27.** 28 × $9.25 _____

PROBLEM SOLVING

23. Twenty-one games were bought for Children's
Hospital. Each game cost $6.98. Was more
than $175.00 spent on the games? _____

Use with Lesson 4-12, text pages 148–149.

Multiplying by Two-Digit Numbers

$21 \times 23 = \underline{?}$

Multiply by the ones.	Multiply by the tens.	Add the partial products.
23	23	23
×21	×21	×21
23	23	23 ← partial
	460	460 ← products
		483

Estimate. Then multiply.

1.   44
 × 22

2.   13
 × 12

3.   21
 × 41

4.   32
 × 12

5.   42
 × 21

6.   43
 × 11

7.   24
 × 12

8.   14
 × 22

9.   12
 × 23

10.   32
 × 13

11.   21
 × 32

12.   22
 × 34

13.   23
 × 12

14.   22
 × 13

15.   33
 × 22

Find the product.

16. $3.00
 ×   40

17. $.64
 ×   11

18. $.32
 ×   32

19. $6.00
 ×   30

20. $.42
 ×   12

PROBLEM SOLVING

21. Beth bought 11 erasers. Each cost $.45. How much did she spend altogether? _____

22. Alex sold 30 magazine subscriptions. If the cost of each subscription was $8.00, what was the total amount? _____

Use with Lesson 4-13, text pages 150–151.   45

More Multiplying by Two-Digit Numbers

Name _____

Date _____

$89 \times 34 = \underline{?}$

Multiply by the ones.	Multiply by the tens.	Add the partial products.
$\overset{3}{34}$	$\overset{3}{\overset{\cancel{3}}{34}}$	$\overset{3}{\overset{\cancel{3}}{34}}$
$\times 89$	$\times 89$	$\times 89$
$\boxed{306} \leftarrow 9 \times 34$	306	306
	$\boxed{2720} \leftarrow 80 \times 34$	2720
		3026

Estimate. Then multiply.

1. 79
$\times 53$

2. 41
$\times 48$

3. 23
$\times 36$

4. 83
$\times 65$

5. 67
$\times 42$

6. 74
$\times 88$

7. 55
$\times 27$

8. 45
$\times 75$

9. 82
$\times 56$

10. 81
$\times 68$

11. $.93
$\times \ 98$

12. $.56
$\times \ 73$

13. $.89
$\times \ 25$

14. $.92
$\times \ 46$

15. $.45
$\times \ 85$

16. $76 \times 39 = $ _____

17. $51 \times \$.37 = $ _____

18. $64 \times \$.48 = $ _____

19. $78 \times 25 = $ _____

PROBLEM SOLVING

20. If one yard equals 36 inches, find the number of inches in 56 yards.

_____

21. The factors are 97 and 78. What is the product?

_____

22. Mary Pat put 45 books on each of 12 shelves. How many books did she put on shelves?

_____

Multiplying with Three-Digit Numbers

Name _____

Date _____

$75 \times 494 = \underline{?}$

Multiply by the ones.	Multiply by the tens.	Add the partial products.
4 2	6 2	6 2
	4̸ 2̸	4̸ 2̸
494	494	494
$\times$ 75	$\times$ 75	$\times$ 75
2 4 7 0	2 4 7 0	2 4 7 0
	3 4 5 8 0	3 4 5 8 0
		3 7,0 5 0

Estimate. Then find the product.

1. 364 $\times$ 45	**2.** 951 $\times$ 79	**3.** 923 $\times$ 37	**4.** 584 $\times$ 67	**5.** 746 $\times$ 53
6. 467 $\times$ 54	**7.** 286 $\times$ 32	**8.** $3.98 $\times$ 62	**9.** $8.59 $\times$ 76	**10.** $6.32 $\times$ 99

11. $56 \times 403 =$ _____

12. $43 \times \$7.38 =$ _____

13. $82 \times \$2.55 =$ _____

14. $28 \times 802 =$ _____

15. $89 \times 619 =$ _____

16. $39 \times 847 =$ _____

17. $63 \times 146 =$ _____

18. $22 \times 591 =$ _____

19. $17 \times 458 =$ _____

20. $94 \times \$3.89 =$ _____

PROBLEM SOLVING

21. The factors are 596 and 48. What is the product? _____

22. The factors are 325 and 16. What is the product? _____

Problem-Solving Strategy: Working Backwards

Name _____

Date _____

Adam got off the bus at 4:00 after visiting a friend.
He spent 30 minutes on the bus each way. He stayed
at his friend's house 1 hour and 15 minutes.
What time did he leave home?

Think: Count back to subtract each time that was added.
4:00 − 30 minutes − 30 minutes − 1 hour 15 minutes = 1:45

Adam left home at 1:45.

Solve. Do your work on a separate sheet of paper.

1. Ms. Sanchez arrived at school at
   7:30 A.M. She spent 35 minutes
   having breakfast at a diner with a
   friend. The trip to school from the
   diner was 15 minutes. What time did
   she leave the diner?

2. Nancy drove 470 miles starting on
   Monday. She drove 90 miles on
   Tuesday and twice as far on
   Wednesday. On Thursday she drove
   100 miles. How far did she drive
   on Monday?

3. Michael bought a shirt for $19 and
   some pants for $18. He received $3
   change. How much money did he
   give the clerk?

4. Maria gave 8 shells to Tanya and 3
   times that many to Liv. She has 50
   shells left. How many shells did she
   have to start with?

5. Stephen played basketball for three
   months and made 73 baskets. He
   made some baskets in January. In
   February he made 20 baskets, and in
   March he made twice that number.
   How many did he make in January?

6. Marcia is 2 years older than Rosemary.
   Rosemary is 3 years older than Lisa.
   Lisa is 2 years younger than her 11-
   year-old brother. How old is Marcia?

Use with Lesson 4-17, text pages 158–159.   Copyright © William H. Sadlier, Inc. All rights reserved.

Division Concepts

Name _____

Date _____

$$3 \div 1 = 3 \qquad 3 \div 3 = 1 \qquad 0 \div 6 = 0$$

$$1\overline{)3}^{\,3} \qquad\qquad 3\overline{)3}^{\,1} \qquad\qquad 6\overline{)0}^{\,0}$$

X X X X X
X X X X X
X X X X X

It is *impossible* to divide a number by 0.

_?_ in each group ⟶ $15 \div 3 = 5$
_?_ equal groups ⟶ $15 \div 5 = 3$

Divide.

1. $5\overline{)5}$ **2.** $3\overline{)0}$ **3.** $1\overline{)8}$ **4.** $8\overline{)8}$ **5.** $8\overline{)0}$

6. $7\overline{)7}$ **7.** $6\overline{)0}$ **8.** $1\overline{)1}$ **9.** $1\overline{)3}$ **10.** $2\overline{)2}$

Find the quotient.

11. $0 \div 5 =$ ____   **12.** $3 \div 3 =$ ____   **13.** $6 \div 6 =$ ____

14. $5 \div 1 =$ ____   **15.** $7 \div 1 =$ ____   **16.** $0 \div 9 =$ ____

17. $4 \div 4 =$ ____   **18.** $9 \div 1 =$ ____   **19.** $8 \div 8 =$ ____

PROBLEM SOLVING

20. The quotient is 8. The dividend is 8. What is the divisor?

_____

21. The divisor is 2. The dividend is 0. What is the quotient?

_____

22. There are 36 cassette tapes. There are 4 cassettes in each package. How many packages are there?

_____

23. Peter has 20 broccoli plants. If he plants them equally in 4 rows, how many plants will be in each row?

_____

24. The divisor is 9. The quotient is 1. What is the dividend?

_____

25. Mario has 42 roses. He puts the same number into each of 7 vases. How many roses are in each vase?

_____

Use with Lesson 5-1, text pages 166–167.

Missing Numbers in Division

Name _____

Date _____

$20 \div \underline{?} = 5$

Think: $5 \times \underline{?} = 20$
$5 \times 4 = 20$

So $20 \div 4 = 5$.

$4\overline{)?}^{\,5}$

Think: $5 \times 4 = \underline{?}$
$5 \times 4 = 20$

So $4\overline{)20}^{\,5}$.

Find the missing divisor.

1. $?\overline{)14}^{\,7}$ 2. $?\overline{)35}^{\,7}$ 3. $?\overline{)40}^{\,5}$ 4. $?\overline{)42}^{\,6}$ 5. $?\overline{)64}^{\,8}$

6. $?\overline{)28}^{\,4}$ 7. $?\overline{)27}^{\,9}$ 8. $?\overline{)16}^{\,2}$ 9. $?\overline{)18}^{\,6}$ 10. $?\overline{)45}^{\,9}$

Find the missing dividend.

11. $7\overline{)?}^{\,3}$ 12. $6\overline{)?}^{\,8}$ 13. $6\overline{)?}^{\,6}$ 11. $9\overline{)?}^{\,2}$ 11. $3\overline{)?}^{\,8}$

16. $7\overline{)?}^{\,8}$ 17. $5\overline{)?}^{\,0}$ 18. $4\overline{)?}^{\,9}$ 16. $5\overline{)?}^{\,5}$ 16. $8\overline{)?}^{\,1}$

Find the missing number.

21. $45 \div \underline{?} = 5$ ____ 22. $56 \div \underline{?} = 8$ ____ 23. $72 \div \underline{?} = 8$ ____

24. $\underline{?} \div 7 = 9$ ____ 25. $\underline{?} \div 4 = 8$ ____ 26. $\underline{?} \div 4 = 9$ ____

PROBLEM SOLVING

27. Bill gave pencils to 4 friends. Each friend got 7 pencils. How many pencils did he have to begin with? _____

28. Frank had 54 books. He put 9 books on each shelf. How many shelves did he use? _____

Number Patterns

What are the next two numbers in the pattern?

2, 5, 5, 8, 8, 11, 11, ?, ?

+3 ×1 +3 ×1 +3 ×1 +3 ×1

Rule: Add 3. Multiply by 1.

The next two numbers in the pattern are 14, 14.

Write the rule for each pattern. Then write the next number.

1. 4, 8, 12, 16, _____

2. 18, 15, 12, 9, _____

3. 6, 12, 18, 24, _____

4. 27, 9, 3, _____

5. 3, 8, 6, 11, 9, _____

6. 12, 11, 28, 27, 44, 43, 60, _____

7. 4, 6, 5, 7, 6, _____

8. 12, 6, 16, 8, 18, _____

9. 2, 4, 5, 10, 11, _____

10. 17, 25, 22, 30, 27, _____

11. 5, 10, 6, 12, 8, 16, _____

12. 16, 23, 25, 32, 34, 41, _____

Write a pattern of 8 numbers for each rule.

13. Rule: Multiply by 3. _____

14. Rule: Add 1. Add 4. _____

15. Rule: Add 20. Subtract 1. _____

Use with Lesson 5-3, text pages 170–171. 51

Estimating in Division

Name _____

Date _____

Estimate: 2268 ÷ 7
Find where the quotient begins.

Think: $7\overline{)2268}$ 7 > 2   Not enough thousands.
$7\overline{)2268}$ 7 < 22 Enough hundreds.

$\dfrac{3}{7\overline{)2268}}$ ← The quotient begins in the hundreds place.

about 300 ← Write zeros for
$7\overline{)2268}$ the other digits.

Put an x in the place where the quotient begins.

1. $3\overline{)92}$ **2.** $7\overline{)42}$ **3.** $5\overline{)642}$ **4.** $8\overline{)805}$

5. $2\overline{)621}$ **6.** $4\overline{)392}$ **7.** $6\overline{)714}$ **8.** $9\overline{)724}$

9. $3\overline{)1267}$ **10.** $5\overline{)2580}$ **11.** $7\overline{)2875}$ **12.** $6\overline{)5460}$

Estimate the quotient.

13. $9\overline{)285}$ **14.** $8\overline{)411}$ **15.** $7\overline{)150}$ **16.** $2\overline{)407}$

17. $4\overline{)2917}$ **18.** $5\overline{)3296}$ **19.** $3\overline{)1201}$ **20.** $6\overline{)567}$

21. $3\overline{)\$3.47}$ **22.** $5\overline{)\$46.11}$ **23.** $7\overline{)\$56.85}$ **24.** $9\overline{)\$81.23}$

25. $7\overline{)\$8.64}$ **26.** $3\overline{)\$8.29}$ **27.** $8\overline{)\$75.25}$ **28.** $5\overline{)\$26.95}$

PROBLEM SOLVING

29. Pat has 1490 nails. If he divides
them equally into 3 packages, about
how many nails will each contain? _____

30. There are 57 marshmallows for roasting.
If 9 children share them equally, about how
many marshmallows will each child have? _____

One-Digit Quotients

Name _____

Date _____

$30 \div 7 = \underline{?}$ Estimate: About how many 7s in 30?

Divide.	Multiply.	Subtract and compare.	Write the remainder.
$\begin{array}{r} 4 \\ 7\overline{)30} \end{array}$	$\times\begin{array}{r} 4 \\ 7\overline{)30} \\ 28 \end{array}$	$\begin{array}{r} 4 \\ 7\overline{)30} \\ -28 \\ \hline 2 \end{array}$ ← 2<7	$\begin{array}{r} 4\ \text{R}2 \\ 7\overline{)30} \\ -28 \\ \hline 2 \end{array}$

Divide.

1. $8\overline{)42}$ **2.** $9\overline{)35}$ **3.** $6\overline{)25}$ **4.** $7\overline{)47}$

5. $6\overline{)31}$ **6.** $4\overline{)18}$ **7.** $5\overline{)27}$ **8.** $7\overline{)46}$

9. $3\overline{)13}$ **10.** $9\overline{)67}$ **11.** $2\overline{)17}$ **12.** $5\overline{)23}$

13. $25 \div 3 = $ _____ **14.** $58 \div 9 = $ _____ **15.** $44 \div 6 = $ _____

16. $44 \div 7 = $ _____ **17.** $32 \div 5 = $ _____ **18.** $29 \div 8 - $ _____

PROBLEM SOLVING

19. Tom has 17 slices of ham. If he uses
3 slices for each sandwich, how many
sandwiches can he make? How many
slices will be left over?   _____

20. Three tennis balls fit into a can. Sancha has
19 tennis balls. How many cans can she fill?
How many tennis balls will be left over?   _____

Divisibility

A number is **divisible** by another number when the remainder is zero.

Divisible by:	100	35	36
2	yes	no	yes
5	yes	yes	no
10	yes	no	no
3	no	no	yes

70	38	47	60	95	88
125	260	135	420	340	632
261	590	612	6621	7640	1245
1110	2450	30,765	930	1600	23,541

1. Write the numbers from the table that are divisible by 2.

_____ _____ _____ _____ _____ _____

_____ _____ _____ _____ _____ _____

2. Write the numbers from the table that are divisible by 5.

_____ _____ _____ _____ _____ _____

_____ _____ _____ _____ _____ _____

3. Write the numbers from the table that are divisible by 10.

_____ _____ _____ _____ _____ _____

_____ _____ _____ _____ _____ _____

4. Write the numbers from the table that are divisible by 3.

_____ _____ _____ _____ _____ _____

_____ _____ _____ _____ _____ _____

Use with Lesson 5-6, text pages 176–177.

Two-Digit Quotients

Name _____

Date _____

$$\begin{array}{r} 2 \\ 3\overline{)75} \\ -6 \\ \hline 15 \end{array}$$

$$\begin{array}{r} 25 \\ 3\overline{)75} \\ -6 \\ \hline 15 \\ -15 \\ \hline 0 \end{array}$$

$$\begin{array}{r} 25 \\ \times3 \\ \hline 75 \end{array}$$

Divide and check.

1. $3\overline{)81}$ 2. $7\overline{)84}$ 3. $2\overline{)94}$ 4. $5\overline{)75}$ 5. $6\overline{)96}$

6. $4\overline{)64}$ 7. $9\overline{)99}$ 8. $8\overline{)96}$ 9. $7\overline{)91}$ 10. $5\overline{)85}$

11. $42 \div 2 =$ _____   12. $96 \div 4 =$ _____

13. $65 \div 5 =$ _____   14. $58 \div 2 =$ _____

15. $68 \div 4 =$ _____   16. $72 \div 3 =$ _____

PROBLEM SOLVING

17. MeiLing cuts a 72-inch long rope into 2 equal parts. How long is each part? _____

18. There are 3 cookies on each plate and 48 cookies in all. How many plates are there? _____

More Two-Digit Quotients

Name _____

Date _____

Divide the tens.	Divide the ones.	Check.

Divide the tens.

```
  ×
  ↑2
4)8 7
 −8 ↓
   7
```

Divide the ones.

```
  ×↓ 2 1  R 3
 4)8 7
  −8 ↓
    7
   −4
    3
```

Check.

```
    2 1
  ×   4
    8 4
  +   3
    8 7
```

Estimate. Then divide and check.

1. 4)9 1   **2.** 3)5 8   **3.** 2)6 5   **4.** 5)8 9   **5.** 6)8 7

6. 3)8 3   **7.** 5)9 2   **8.** 7)9 9   **9.** 8)9 1   **10.** 2)7 3

11. 49 ÷ 3 = _____   **12.** 67 ÷ 4 = _____

13. 59 ÷ 2 = _____   **14.** 83 ÷ 5 = _____

15. 74 ÷ 6 = _____   **16.** 99 ÷ 2 = _____

PROBLEM SOLVING

17. Yusuf divides 38 comic books equally among 4 friends. How many comic books does each friend receive? How many comic books are left over?

_____

Three-Digit Quotients

Name _____

Date _____

Division Steps

1. Estimate.	2. Divide.	3. Multiply.	4. Subtract.
5. Compare.	6. Bring Down.	7. Repeat steps as necessary.	8. Check.

Complete.

1.
```
      1 3 4
  7)9 3 8
   -7↓
   □□
  -2 1↓
     2 8
    -□□
       0
```

Check.
```
  □□□
×     7
  9 3 8
```

2.
```
      1 3 8  R 3
  4)5 5 5
   -4↓
    1 5
   -1 2↓
      3 5
     -□□
       □
```

Check.
```
    1 3 8
×       4
    5 5 2
+       □
    5 5 5
```

Estimate. Then divide and check.

3. 5)612   4. 3)594   5. 2)912   6. 5)623

7. 4)953   8. 6)772   9. 3)944   10. 6)826

11. 7)861   12. 5)810   13. 8)978   14. 3)372

More Difficult Quotients

Name _____

Date _____

$6\overline{)524}$

Think: $\mathbf{6}\overline{)524}$ $6 > 5$ Not enough hundreds.
$\mathbf{6}\overline{)\mathbf{52}4}$ $6 < 52$ Enough tens.

$$
\begin{array}{r}
87 \ \ \text{R}\,2 \\
6\overline{)524} \\
-48\downarrow \\
\hline
44 \\
-42 \\
\hline
2
\end{array}
$$

Check:
$$
\begin{array}{r}
87 \\
\times\ \ 6 \\
\hline
522 \\
+\ \ 2 \\
\hline
524
\end{array}
$$

Estimate. Then find the quotient.

1. $6\overline{)440}$ **2.** $7\overline{)597}$ **3.** $9\overline{)703}$ **4.** $6\overline{)576}$

5. $5\overline{)448}$ **6.** $8\overline{)764}$ **7.** $3\overline{)136}$ **8.** $7\overline{)185}$

9. $4\overline{)356}$ **10.** $5\overline{)327}$ **11.** $7\overline{)301}$ **12.** $9\overline{)563}$

PROBLEM SOLVING

13. Seven friends shared 259 stamps equally. How many did each friend receive? _____

14. A basket holds 9 quarts of berries. How many baskets will 345 quarts fill? How many quarts will be left over? _____

15. A farmer delivered 590 bales of hay. Each of 6 barns received the same number of bales. At most, how many bales did each get? How many were left over? _____

Use with Lesson 5-10, text pages 184–185.

Zeros in the Quotient

$$4\overline{)836}$$

Think: $4 < 8$
Enough hundreds.

```
    2 0 9
4) 8 3 6
  - 8
   0 3  ← 3 < 4 Not enough tens.
  - 0      Write 0 in the tens place.
    3 6
  - 3 6
    0
```

Check:
```
  209
×   4
  836
```

Divide.

1. $3\overline{)912}$　　　　**2.** $8\overline{)859}$　　　　**3.** $2\overline{)441}$　　　　**4.** $9\overline{)975}$

5. $2\overline{)461}$　　　　**6.** $6\overline{)950}$　　　　**7.** $8\overline{)835}$　　　　**8.** $9\overline{)962}$

9. $4\overline{)683}$　　　　**10.** $9\overline{)981}$　　　　**11.** $7\overline{)734}$　　　　**12.** $6\overline{)631}$

PROBLEM SOLVING

13. The divisor is 3. The dividend is 931.
Find the quotient and remainder. _____

14. The dividend is 614. The divisor is 3.
Find the quotient and remainder. _____

Larger Numbers in Division

Name _____

Date _____

$4828 \div 9 =$ <u>?</u>

```
    536  R 4      Check:
  9)4828            536
   -45↓             ×   9
    32             4824
   -27↓            +   4
    58             4828
   -54
     4
```

$8734 \div 6 =$ <u>?</u>

```
   1455  R 4      Check:
  6)8734           1455
   -6↓             ×   6
    27             8730
   -24↓            +   4
    33             8734
   -30↓
    34
   -30
     4
```

Divide and check.

1. 4)8 1 8 2

2. 3)9 2 1 9

3. 6)9 8 4 2

4. 5)4 6 3 0

5. 8)3 9 1 5

6. 7)6 9 0 4

7. 3)7 5 7 9

8. 9)8 6 2 5

PROBLEM SOLVING

9. A company shipped 8916 toys in 4 weeks. It shipped the same number of toys each week. At most, how many toys did it ship each week?

_____

60   **Use with Lesson 5-12, text pages 188–189.**

Dividing Money

Name _____

Date _____

$65.16 ÷ 9 = __?__

Write the dollar sign and decimal point in the quotient above the dollar sign and decimal point in the dividend.

Divide as usual.

```
   $  7.2 4
9)$ 6 5.1 6
  − 6 3
      2 1
    − 1 8
        3 6
      − 3 6
          0
```

Check:

```
    $7.2 4
  ×       9
  $6 5.1 6
```

Divide and check.

1. 5)$1.3 0

2. 9)$9.5 4

3. 2)$8.3 6

4. 8)$5.3 6

5. 9)$1 1.4 3

6. 5)$6 7.9 5

7. 5)$4 1.6 5

8. 7)$8 7.6 4

PROBLEM SOLVING

9. Mr. Akham sold 5 identical bird feeders for $74.85. What was the cost of one bird feeder?

_____

10. All birdhouses are on sale for the same price. Mr. Akham sold 9 of them for $89.82. What was the sale price for each birdhouse?

_____

Order of Operations

Name _____

Date _____

Simplify: $3 \times 2 + 10 - 20 \div 5 =$ __?__

First multiply or divide. Work in order from left to right.
Then add or subtract.

$$3 \times 2 + 10 - 20 \div 5 = \quad 6 + 10 - 4 = 12$$

Use the order of operations to simplify.

1. $13 + 8 - 5 =$ _____

2. $10 - 4 + 7 =$ _____

3. $7 \times 6 - 2 =$ _____

4. $46 - 4 \times 9 =$ _____

5. $8 \times 9 - 16 =$ _____

6. $30 + 20 - 4 =$ _____

7. $12 - 4 + 6 \times 3 =$ _____

8. $12 - 8 \div 4 + 6 =$ _____

9. $8 \times 6 + 24 \div 8 =$ _____

10. $6 \times 4 - 12 \div 2 =$ _____

11. $20 + 6 \div 6 - 7 =$ _____

12. $45 - 10 + 81 \div 9 =$ _____

13. $200 \div 4 - 50 =$ _____

14. $50 + 100 \div 2 \div 10 =$ _____

15. $40 \times 2 + 10 - 45 =$ _____

16. $18 \times 2 \div 9 + 4 =$ _____

17. $6 + 5 \div 1 - 3 \times 3 =$ _____

18. $25 \div 5 + 4 \times 3 - 3 =$ _____

19. $23 - 21 \div 7 + 5 \times 2 =$ _____

20. $68 + 10 \div 2 - 7 \times 2 =$ _____

21. $64 \div 8 \times 2 + 50 + 37 =$ _____

22. $6 \times 3 \div 9 + 100 - 102 =$ _____

Use with Lesson 5-14, text pages 192–193.

Finding Averages

Find the average: 336, 118, 209

Add:
```
    336
    118
  + 209
  ─────
    663
```

Divide the sum by the number of addends.

```
       221 ←── average
   3)663
```

The average is 221.

Find the average.

1. 55, 91, 76

_____

2. 86, 50, 35

_____

3. 23, 31, 19, 47

_____

4. 253, 147, 383

_____

5. 29, 42, 53, 40

_____

6. 45, 238, 70, 875

_____

7. $2.15, $1.98, $1.00, $3.79

_____

8. $3.05, $2.29, $1.30, $4.04

_____

PROBLEM SOLVING

9. In 5 days, Mike's Gas Station serviced 20, 45, 100, 85, and 60 cars. What was the average number of cars serviced each day? _____

Problem-Solving Strategy: Interpret the Remainder

Name _____

Date _____

In the morning, 196 rolls were delivered to a cafeteria. Each person was served 3 rolls. How many more rolls were needed to serve an extra person?

$$
\begin{array}{r}
65 \text{ R } 1 \\
3 \overline{)196} \\
-18 \\
\hline
16 \\
-15 \\
\hline
1
\end{array}
$$

$3 - 1 = 2$

Two more rolls were needed.

Solve. Do your work on a separate sheet of paper.

1. Pizzas are be cut into 8 slices. How many pizzas are needed to serve one slice to each of 185 people?

2. Each pitcher of punch can serve 9 people. How many pitchers are needed to serve 232 people?

3. A clerk at Oodles of Rings put 6 rings in each display case. How many cases are needed for 200 rings?

4. Napkins came in a box of 500. How many people could use 3 napkins each? How many more would be needed for another person?

5. Each table can seat 7. What is the least number of tables needed to seat 155 people?

6. Mandy made salads. There were 3 tomato slices on each. How many salads did she make with 121 slices?

7. People sat on benches for a concert. Each bench could seat 7 people. What was the least number of benches needed to seat 192 people?

8. Relay races called for 8 people on a team. There were 43 men and 34 women who wanted to race. How many teams could be formed? How many more people were needed for another team?

Measuring with Inches

Name _____

Date _____

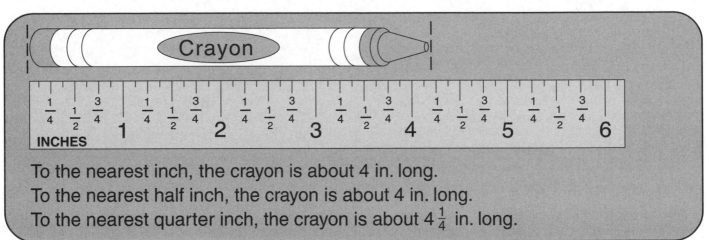

To the nearest inch, the crayon is about 4 in. long.
To the nearest half inch, the crayon is about 4 in. long.
To the nearest quarter inch, the crayon is about $4\frac{1}{4}$ in. long.

Measure each to the nearest inch, nearest half inch, and nearest quarter inch.

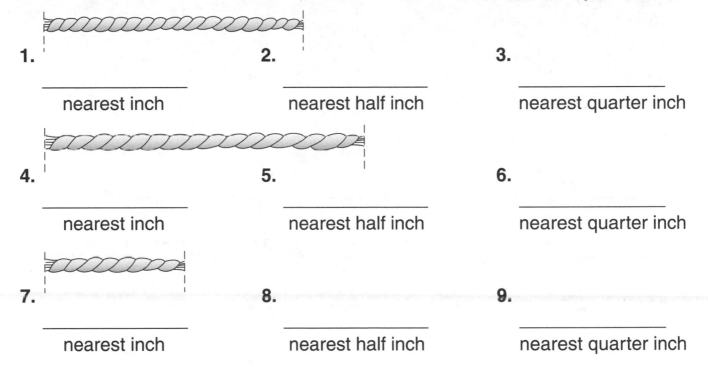

1.   **2.**   **3.**

_____   _____   _____
nearest inch   nearest half inch   nearest quarter inch

4.   **5.**   **6.**

_____   _____   _____
nearest inch   nearest half inch   nearest quarter inch

7.   **8.**   **9.**

_____   _____   _____
nearest inch   nearest half inch   nearest quarter inch

Draw a line segment for each length.

10. 2 in.

11. $4\frac{1}{2}$ in.

12. $3\frac{3}{4}$ in.

13. $5\frac{1}{2}$ in.

Renaming
Units of Length

Compare: 72 in. __?__ 3 yd

Make a table.

in.	36	72	108
yd	1	2	3

So 72 in. < 3 yd.

Multiply to rename the larger unit.

3 yd = (3 × 36) in.
3 yd = 108 in.

Divide to rename the smaller unit.

72 in. = (72 ÷ 36) yd
72 in. = 2 yd

Compare. Write <, =, or >.
You may make a table or compute.

1. 48 in. _____ 3 ft

2. 15 ft _____ 6 yd

3. 4 yd _____ 12 ft

4. 144 in. _____ 7 yd

5. 8 ft _____ 120 in.

6. 7 yd _____ 20 ft

7. 25 ft _____ 8 yd

8. 72 in. _____ 6 ft

9. 48 in. _____ 5 ft

10. 18 ft _____ 7 yd

11. 6 yd _____ 210 in.

12. 3 mi _____ 15,840 ft

13. 5280 yd _____ 4 mi

14. 10,000 ft _____ 2 mi

PROBLEM SOLVING

15. Pedro made a domino chain that was
132 in. long. Muni made one that
was 16 feet long. Whose chain was
longer? by how many inches?

_____

Computing Customary Units

Name _____

Date _____

Add or subtract the smaller units first.
Rename units as needed.

$$
\begin{array}{r}
3\text{ ft }\ 6\text{ in.} \\
+\ 2\text{ ft }\ 8\text{ in.} \\
\hline
5\text{ ft 14 in.} = 5\text{ ft} + 1\text{ ft} + 2\text{ in.} = 6\text{ ft 2 in.}
\end{array}
$$

$$
\begin{array}{r}
6\text{ yd 2 ft} \\
-\ 1\text{ yd 1 ft} \\
\hline
5\text{ yd 1 ft}
\end{array}
$$

Add.

1. 5 ft 7 in.
\+ 6 ft 2 in.

2. 7 yd 1 ft
\+ 4 yd 1 ft

3. 18 ft 5 in.
\+ 27 ft 4 in.

4. 4 yd 2 ft
\+3 yd 2 ft

5. 1 ft 9 in.
\+4 ft 5 in.

6. 2 yd 3 ft
\+ 3 yd 2 ft

7. 2 ft 6 in.
\+ 3 ft 7 in.

8. 3 yd 2 ft
\+   2 ft

9. 4 yd 1 ft + 3 yd = _____

10. 4 ft 9 in. + 3 ft 7 in. = _____

Subtract.

11. 15 ft 7 in.
− 9 ft 5 in.

12. 28 yd 10 ft
−19 yd   6 ft

13. 72 yd 2 ft
− 48 yd 1 ft

14. 16 ft 11 in.
−16 ft   8 in.

15. 7 ft 9 in.
−   6 in.

16. 14 ft 6 in.
− 9 ft 2 in.

17. 5 yd 2 ft
− 3 yd

18. 4 yd 2 ft
−1 yd 1 ft

19. 7 ft 10 in. − 3 in. = _____

20. 9 yd 2 ft − 2 ft = _____

PROBLEM SOLVING

21. It takes 2 yd 1 ft of material to make a certain
type of skirt. How much material is needed
for two of these skirts?   _____

22. Myron cut a section 8 ft 7 in. long from a board
that was 12 ft 11 in. long. How long was the
piece of board that was left?   _____

Customary Units of Capacity

Name _____

Date _____

1 fl oz	1 c	1 pt	1 qt	1 gal
8 fluid ounces = 1 cup	2 cups = 1 pint	2 pints = 1 quart	4 quarts = 1 gallon	

Complete each table.

1.

gal		2	3		5
qt	4			16	
pt	8	16			40

2.

pt	1				5
c			4	6	10
fl oz	16	32			

Compare. Write <, =, or >. You may make a table or compute.

3. 6 pt _____ 2 qt

4. 6 gal _____ 50 qt

5. 6 qt _____ 2 gal

6. 11 c _____ 86 fl oz

7. 20 qt _____ 10 gal

8. 8 pt _____ 8 c

9. 7 qt _____ 15 pt

10. 5 c _____ 92 fl oz

11. 75 fl oz _____ 10 c

PROBLEM SOLVING

12. Shawna drinks 56 fluid ounces of water each day. Does she drink 8 cups of water each day?

_____

13. Luis sold 10 one-gal cartons of milk and 36 one-qt cartons of milk. Did he sell more milk in gallon or quart cartons?

_____

14. Sue's guests finished 2 gal of punch. If each guest drank 1 c of punch, how many guests did she have?

_____

Customary Units of Weight

1 oz   **1 lb**   **1 T**

16 ounces = 1 pound   2000 pounds = 1 ton

Write *oz*, *lb*, or *T* for the unit you would use to measure each.

1. _____

2. _____

3. _____

4. _____

5. _____

6. _____

Compare. Write <, =, or >.

7. 6000 lb _____ 2 T

8. 3 lb _____ 48 oz

9. 8T _____ 12,000 lb

10. 1 T _____ 3000 lb

11. 96 oz _____ 6 lb

12. 54 oz _____ 3 lb

13. 10 oz _____ 1 lb

14. 31 oz _____ 2 lb

15. 18 oz _____ 1 lb

Draw a line to the tool you would use to measure each.

16. length of a notebook   yardstick

17. weight of a dog   measuring cup

18. water for a pitcher   ruler

19. height of a door   scale

PROBLEM SOLVING

20. A tractor trailer weighs 5 tons. How many pounds is this? _____

Measuring with Metric Units

Name _____

Date _____

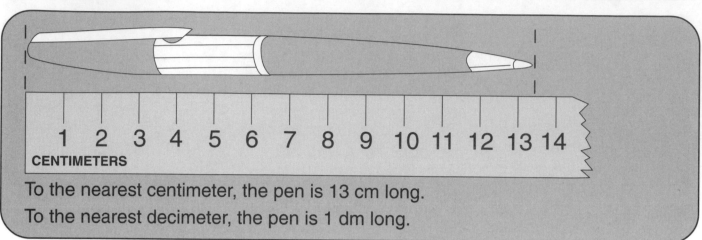

To the nearest centimeter, the pen is 13 cm long.

To the nearest decimeter, the pen is 1 dm long.

Measure each to the nearest centimeter.

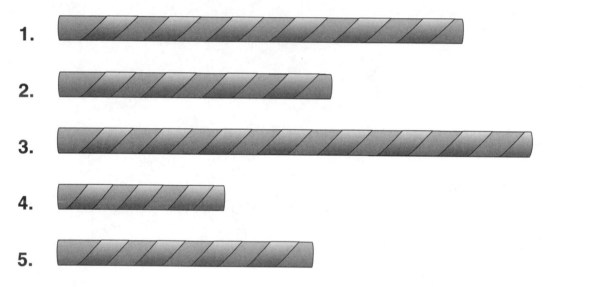

1. _____

2. _____

3. _____

4. _____

5. _____

Draw a line segment for each length.

6. 6 cm

7. 1 dm

8. 3 cm

9. 8 cm

10. 5 cm

Working with Metric Units

Compare: 200 cm __?__ 3 m

Make a table.

cm	100	200	300
m	1	2	3

So 200 cm < 3 m.

Multiply to rename the larger unit.

$3 \text{ m} = (3 \times 100) \text{ cm}$
$3 \text{ m} = 300 \text{ cm}$

Divide to rename the smaller unit.

$200 \text{ cm} = (200 \div 100) \text{ m}$
$200 \text{ cm} = 2 \text{ m}$

Compare. Write <, =, or >.
You may make a table or compute.

1. 12 dm _____ 1m

2. 4 cm _____ 500 mm

3. 2000 m _____ 2 km

4. 15 dm _____ 1500 cm

5. 600 mm _____ 7 dm

6. 80 m _____ 8000 cm

7. 3 km _____ 30 000 m

8. 1200 cm _____ 10 m

9. 600 cm _____ 5 dm

10. 9000 m _____ 9 km

11. 3000 dm _____ 30 m

12. 40 cm _____ 4000 mm

13. 10 000 m _____ 100 km

14. 20 km _____ 200 dm

PROBLEM SOLVING

15. Write these lengths in order from shortest to longest: 200 m, 20 cm, 20 km, 200 dm, 2 000 000 mm.

_____

Metric Units of Capacity

About 20 drops of liquid equal 1 milliliter.

1000 milliliters (mL) = 1 liter (L)

Circle the letter of the best estimate.

1. bottle of seltzer   **a.** 1 mL   **b.** 10 mL   **c.** 1 L

2. bucket of water   **a.** 800 mL   **b.** 8 L   **c.** 80 L

3. glass of juice   **a.** 18 mL   **b.** 180 mL   **c.** 1800 mL

4. cup of milk   **a.** 240 mL   **b.** 24 L   **c.** 24 mL

Write *mL* or *L* for the unit you would use to measure the capacity of each.

5. cup of punch _____

6. aquarium _____

7. swimming pool _____

8. medicine dropper _____

Compare. Write <, =, or >.
You may make a table or compute.

9. 700 mL ____ 7 L

10. 6 L ____ 5500 mL

11. 2000 mL ____ 20 L

12. 8000 L ____ 8 mL

13. 150 L ____ 1500 mL

14. 9000 mL ____ 9 L

PROBLEM SOLVING

15. Two thermoses each hold 720 mL of liquid. If both are filled, will they together hold more or less than 1 L?   _____

16. Which holds more: a bowl with a 5000 mL capacity or one with a 4 L capacity?   _____

Metric Units of Mass

Name _____

Date _____

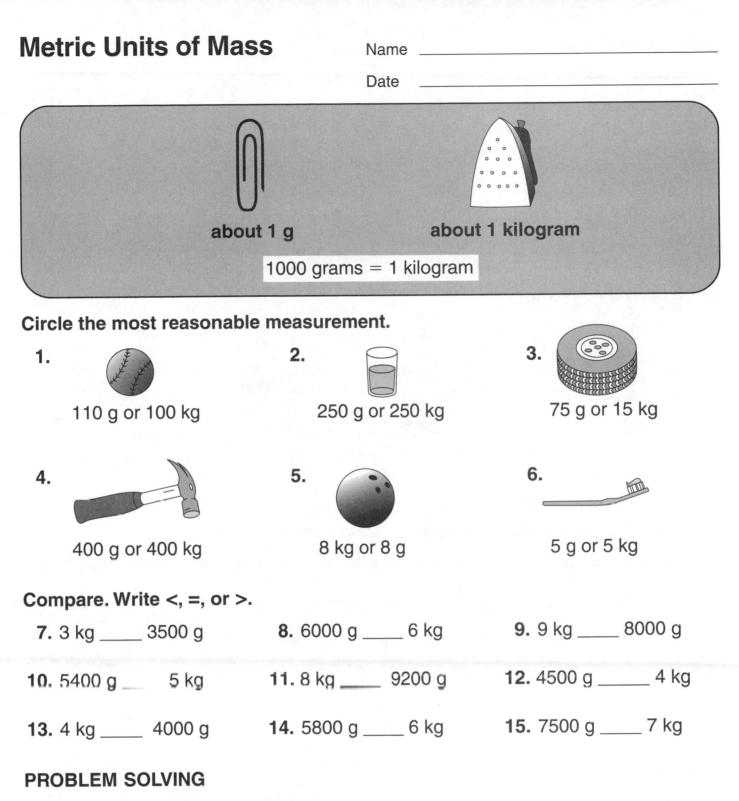

about 1 g

about 1 kilogram

1000 grams = 1 kilogram

Circle the most reasonable measurement.

1. 110 g or 100 kg

2. 250 g or 250 kg

3. 75 g or 15 kg

4. 400 g or 400 kg

5. 8 kg or 8 g

6. 5 g or 5 kg

Compare. Write <, =, or >.

7. 3 kg _____ 3500 g

8. 6000 g _____ 6 kg

9. 9 kg _____ 8000 g

10. 5400 g _____ 5 kg

11. 8 kg _____ 9200 g

12. 4500 g _____ 4 kg

13. 4 kg _____ 4000 g

14. 5800 g _____ 6 kg

15. 7500 g _____ 7 kg

PROBLEM SOLVING

16. A hot dog has about 15 g of fat. A slice of pizza has about 19 g of fat. Which food has more fat? How much more? _____

17. A box of raisins has a mass of 425 g. Would two boxes of raisins have a mass that is more or less than a kilogram? _____

Temperature

Name _____

Date _____

Temperature can be measured in **degrees Fahrenheit (°F)**
or **degrees Celsius (°C)**.

Each line on the Fahrenheit scale
stands for 2° F.

Each line on the Celsius scale
stands for 1° C.

Average body temperature is 98.6° F

Average body temperature is 37° C.

Write each temperature.

1.
°F

_____

2.
°F

_____

3.
°C

_____

4.
°C

_____

Ring the letter of the most reasonable temperature.

5. cold lemonade   **a.** 5° C   **b.** 50° C   **c.** 25° C   **d.** 75° C

6. cup of hot tea   **a.** 85° C   **b.** 30° C   **c.** 10° C   **d.** −10° C

7. baking a pie   **a.** 19° C   **b.** 40° C   **c.** 85° C   **d.** 190° C

8. room   **a.** 25° F   **b.** 50° F   **c.** 68° F   **d.** 95° F

9. ice cube   **a.** 30° F   **b.** 40° F   **c.** 65° F   **d.** 80° F

Complete.

10. | Start with 25° C | Rise of 4° C | [ ] | Drop of 6° C | [ ] | Drop of 2° C | [ ] |

PROBLEM SOLVING

11. One day the temperature was 72° F.
The next day it was 65° F. What was the
difference between the two temperatures? _____

Time

Name _____

Date _____

Read: 25 minutes to 9   or 35 minutes past 8
Write: 8:35

Write each time.

1. __ : __

_____ minutes past _____

_____ minutes to _____

2. __ : __

_____ minutes past _____

_____ minutes to _____

3. __ : __

_____ minutes past _____

Match the times.

4. _____ 5 minutes to 9

5. _____ 30 minutes past 1

6. _____ 15 minutes past 2

7. _____ 10 minutes past 3

a. `2:15`

b. `3:10`

c. `8:55`

d. `1:30`

Write A.M. or P.M. to make each statement reasonable.

8. Jamal eats lunch at 12:30 _____.

9. The store opens at 9:30 _____.

Write the time. Use A.M. or P.M.

10. 12 minutes past 6 at night _____

11. 15 minutes to midnight _____

12. 26 minutes to 11 in the morning _____

Elapsed Time

Name _____

Date _____

How much time has passed?

A.M.

P.M.

Count the hours by 1s.
Count the minutes by 5s and 1s.

From 9:28 A.M. to 3:28 P.M. ────→ 6 h
From 3:28 P.M. to 3:35 P.M. ────→ 7 min

6 hours and 7 minutes have passed.

Write how much time has passed.

1. from 9:10 A.M. to 9:45 A.M. _____

2. from 10:50 P.M. to 11:23 P.M. _____

3. from 10:30 P.M. to 12:43 A.M. _____

4. from 11:45 A.M. to 3:52 P.M. _____

5. from 11:30 P.M. to 12:10 A.M. _____

6. from 9:31 A.M. to 11:45 A.M. _____

7. from 11:06 A.M. to 7:52 P.M. _____

8. P.M. `11:45`   A.M. `2:27`

_____

9. P.M.   P.M.

_____

PROBLEM SOLVING

10. A plane leaves Miami at 12:09 P.M. and lands in New York at 2:12 P.M. The flight was supposed to take 1 hour 52 minutes. Was the plane on time?

_____

11. The date is August 17. What is the date 1 week before? 13 days before? 2 weeks after?

_____

Use with Lesson 6-12, text pages 228–229.

Problem-Solving Strategy: Two-Step Problems

Name _____

Date _____

> Edward works at Pete's Pizza Place. He used 3 lb
> of white flour and two 24-oz bags of wheat flour.
> How much flour did he use altogether?
>
> Add to find the amount
> of wheat flour: 24 oz + 24 oz = 48 oz or 3 lb
>
> Add to find the total: 3 lb + 3 lb = 6 lb
>
> He used 7 lb of flour altogether.

Solve. Do your work on a separate sheet of paper.

1. Pete sells a Party Pleaser pizza. He uses 4 oz of cheddar cheese and twice that amount of mozzarella cheese. How much cheese does he use in all?

   _____

2. Pete made 12 sausage pizzas and 36 cheese pizzas. Mario made 57 pizzas in all. How many more pizzas did Mario make than Pete?

   _____

3. A customer bought 2 small pizzas that cost $5.95 each. How much was her change from $15.00?

   _____

4. Phillip bought 1 medium pizza that cost $8.50 and a drink that cost $0.79. How much change did he receive from a $20 bill?

   _____

5. On Friday, 29 customers ordered pizzas. On Saturday, twice that many ordered pizzas. How many customers in all ordered pizzas on Friday and Saturday?

   _____

6. A pizza was cut into 16 slices. Oscar ate 5 slices, Zhao ate 3 slices, and Tiffany ate 4 slices. How many slices were left over?

   _____

Graphing Sense

Name _____

Date _____

Complete each graph. Then label each as a *circle graph, pictograph, line graph,* or *bar graph.*

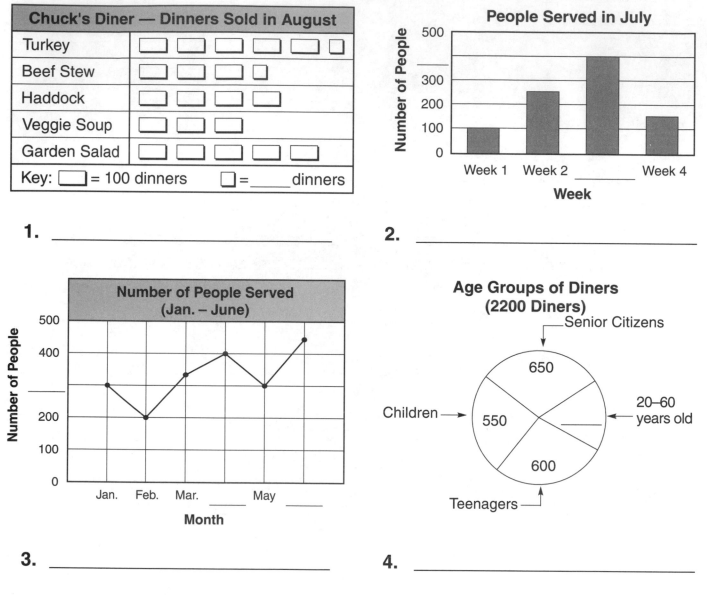

Chuck's Diner — Dinners Sold in August	
Turkey	▭ ▭ ▭ ▭ ▭ ▯
Beef Stew	▭ ▭ ▭ ▯
Haddock	▭ ▭ ▭ ▭
Veggie Soup	▭ ▭ ▭
Garden Salad	▭ ▭ ▭ ▭ ▭
Key: ▭ = 100 dinners ▯ = _____ dinners	

People Served in July

1. _____

2. _____

Number of People Served (Jan. – June)

Age Groups of Diners (2200 Diners)

Senior Citizens — 650
Children → 550
20–60 years old
Teenagers → 600

3. _____

4. _____

PROBLEM SOLVING. Use the graphs above.

5. What is the age group with the greatest number of diners? _____

6. In which month were the fewest number of people served? _____

7. How many more turkey dinners than haddock dinners were sold? _____

8. Which week in July had the greatest number of customers? _____

Making Pictographs

Name _____

Date _____

Favorite Vegetables of Students	
Peas	~~HHT~~ ~~HHT~~ ~~HHT~~ ~~HHT~~ ~~HHT~~ ~~HHT~~ ~~HHT~~ ~~HHT~~
Corn	~~HHT~~ ~~HHT~~ ~~HHT~~ ~~HHT~~ ~~HHT~~ ~~HHT~~ ////
Green Peppers	~~HHT~~ ~~HHT~~ ~~HHT~~ ~~HHT~~
Cucumbers	~~HHT~~ ~~HHT~~ ~~HHT~~ ~~HHT~~ ///
Carrots	~~HHT~~ ~~HHT~~ ~~HHT~~ ~~HHT~~ ~~HHT~~ ~~HHT~~
Squash	~~HHT~~ ~~HHT~~ ~~HHT~~

Make a pictograph. Use the tally chart above.

1.

Favorite Vegetables of Students	
Key:	

2. Write a question for the pictograph you made. _____

PROBLEM SOLVING

Use the pictograph at the right.

3. How many more cheese than mushroom pizzas were sold?

_____

Pizzas Sold in May	
Sausage	◯ ◯ ◯ ◯ ◯
Mushroom	◯ ◯ ◯ ◖
Cheese	◯ ◯ ◯ ◯ ◯ ◯ ◯
Anchovy	◖
Key: ◯ = 20 pizzas ◖ =10 pizzas	

4. Which kind of pizza was least popular?

_____

5. How many sausage pizzas were sold?

_____

6. How many pizzas were sold in May? _____

Making Bar Graphs

Name _____

Date _____

Books Read Last Year			
Name	**Total**	**Name**	**Total**
Brian	60	Edgar	90
Sonia	80	Nykesha	100
Ellen	90	Yushiro	120

Make a bar graph. Use the table above.

1.

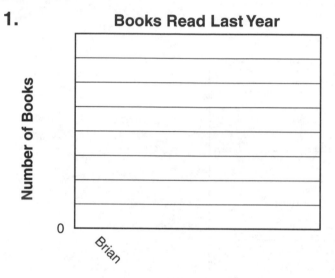

2. Write a question for the bar graph you made.

_____

_____

_____

_____

_____

PROBLEM SOLVING Use the bar graph below.

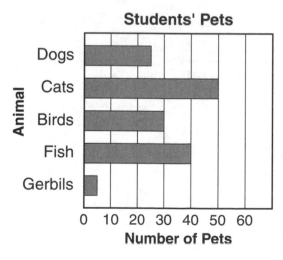

3. How many pets are dogs?

_____

4. How many more cats than gerbils are pets?

_____

5. Which pet is shown by the bar for 30?

_____

6. How many fewer dogs than fish are pets?

_____

Use with Lesson 7-3, text pages 244–245.

Line Graphs

Name _____

Date _____

PROBLEM SOLVING Use the line graph below.

1. In which month did Tony receive a score of 85?

   _____

2. What was the highest score Tony received?

   _____

3. Which score did he receive more than once?

   _____

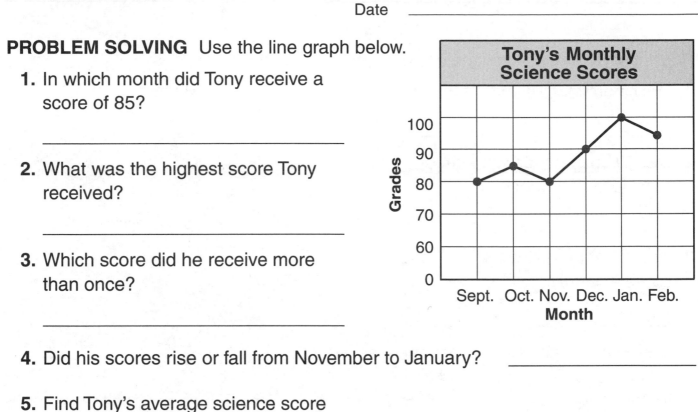

4. Did his scores rise or fall from November to January? _____

5. Find Tony's average science score from October to February. _____

Use the line graph at the right for problems 6–10.

6. In which month was the greatest number of books sold?

   _____

7. During which month were 1500 books sold?

   _____

8. What was the total number of the books sold in March and April?

   _____

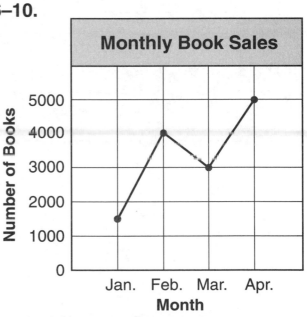

9. Between which two months was the increase in book sales the greatest? _____

10. How many books were sold during all four months? _____

Circle Graphs

Name _____

Date _____

PROBLEM SOLVING Use the circle graph at the right.

1. How much did Jill budget for snacks?

_____

2. Did Jill budget more for school supplies or for snacks?

_____

3. Which items were budgeted for the same amount?

_____

4. During certain weeks, Jill did not need to pay dues. She added that amount to the college fund. How much was in the college fund during those weeks?

_____

Jill's Budget

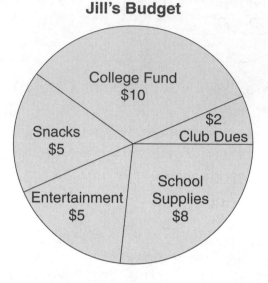

5. How much money does the graph represent in all?

_____

How many of the total number of construction workers were

6. a. carpenters

_____

b. bricklayers

_____

c. electricians and welders

_____

d. plumbers

_____

7. Which group had the greatest number of construction workers?

_____

Number of Construction Workers

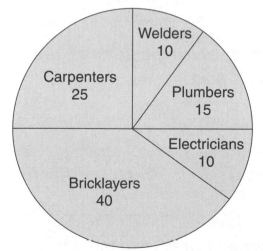

8. Which two groups had the same number of construction workers?

_____

Combinations

Hats: red, white

Scarves: red, white, blue

How many combinations?

Hat	Scarf	Combination
red	red	red, red
	white	red, white
	blue	red, blue
white	red	white, red
	white	white, white
	blue	white, blue

6 combinations   2 × 3 = 6

PROBLEM SOLVING

1. John has a blue shirt, a white shirt, and a gray shirt. He has a striped tie and a flower print tie. How many combinations of shirt and tie can he wear?

_____

2. Lila has opal earrings, pearl earrings, and garnet earrings. She has 1 gold and 1 silver necklace and 1 black and 1 gold watch. How many combinations of earrings, necklace, and watch can she wear?

_____

3. For lunch Kyko can have either soup or salad, either a tuna, a chicken, or a ham sandwich, on either wheat, white, or rye bread. How many combinations can she make?

_____

4. On Monday the cafeteria serves either spaghetti or lasagne with either milk or juice. The dessert is either peaches or pears. How many combinations can be served?

_____

5. On Saturday Kahlid can go to the movie, a ball game, or bowling. Afterward he can have supper at either a pizza, burger, taco, or chicken restaurant. How many ways can he choose to spend Saturday?

_____

6. Sundaes R Us offers a choice of vanilla, chocolate, or coffee ice cream. This can be topped with either hot fudge, butterscotch, or caramel sauce. Decorations can be either chocolate sprinkles, nuts, or toffee bits. How many combinations can be made?

_____

Predicting Probability

Name _____

Date _____

Use the coins.

At random, is it *more likely, equally likely*, or *less likely* that you would pick

1. the half dollar than a quarter? _____

2. a dime or a penny? _____

3. a nickel than a quarter? _____

4. a quarter than a dime? _____

5. a penny than a nickel? _____

6. a nickel than a dime? _____

Use the spinner. What is the probability of the spinner landing on

7. 7? _____

8. 3? _____

9. 1? _____

10. 4? _____

11. not 5? _____

12. 2? _____

13. even? _____

14. odd? _____

Use the set of blocks. At random, what is the probability that you would pick

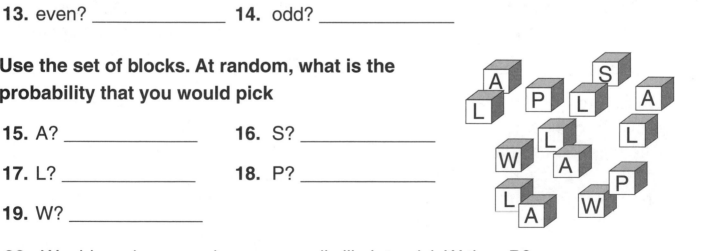

15. A? _____

16. S? _____

17. L? _____

18. P? _____

19. W? _____

20. Would you be *more, less,* or *equally likely* to pick W than P? _____

A than L? _____ P than S? _____

Events and Outcomes

There are four game tokens in a bag: red, yellow, green and blue. What is the probability of picking the red token at random *on the first try?*

Probability of picking red: 1 out of 4.

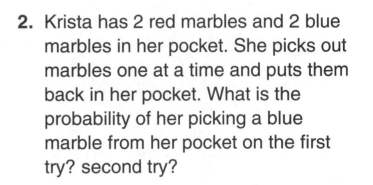

PROBLEM SOLVING

1. Randall wrote the numbers 1, 2, 3, and 4 on 4 slips of paper and put them into a bag. He picked a number and put it back into the bag. What is the probability of picking 2 on the fourth try?

   _____

2. Krista has 2 red marbles and 2 blue marbles in her pocket. She picks out marbles one at a time and puts them back in her pocket. What is the probability of her picking a blue marble from her pocket on the first try? second try?

   _____

3. Jack has a cube with a picture of a different insect on each side: an ant, a beetle, a fly, a bee, a mosquito, and a moth. What is the probability that the ant side will land face up on the 1st roll of the cube? Does the probability change on the 20th try?

   _____

4. Diane wrote each letter of her name on a piece of paper. She picked a piece of paper at random and threw it away. She did this for each try. If D was not picked on the first two tries, what was the probability of picking the D on the third try?

   _____

5. Kim has 3 blue marbles, 3 red marbles, and 2 green marbles in a bag. What is the probability of picking a red marble on the 1st try? a green marble? a blue marble?

   _____

6. Two boys have a box with 3 polished rocks of equal size. Without looking, each boy tries to pick his favorite rock. What is the probability that the boy who picks first will pick his favorite?

   _____

Problem-Solving Strategy: Make Up A Question

Name _____

Date _____

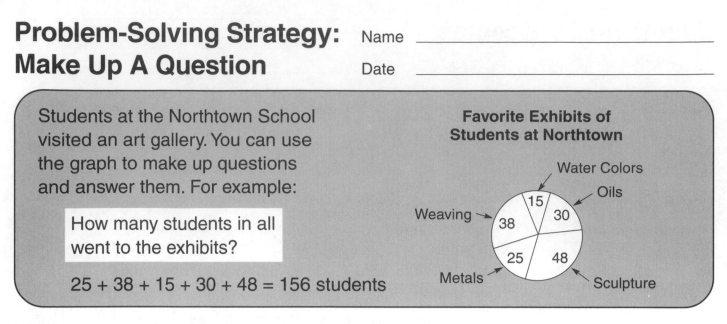

Students at the Northtown School visited an art gallery. You can use the graph to make up questions and answer them. For example:

Favorite Exhibits of Students at Northtown

How many students in all went to the exhibits?

25 + 38 + 15 + 30 + 48 = 156 students

Water Colors
Oils
15
30
Weaving → 38
25   48
Metals   Sculpture

Write one question for the circle graph above. Solve.

1. _____

_____

Write three questions for each graph. Solve.

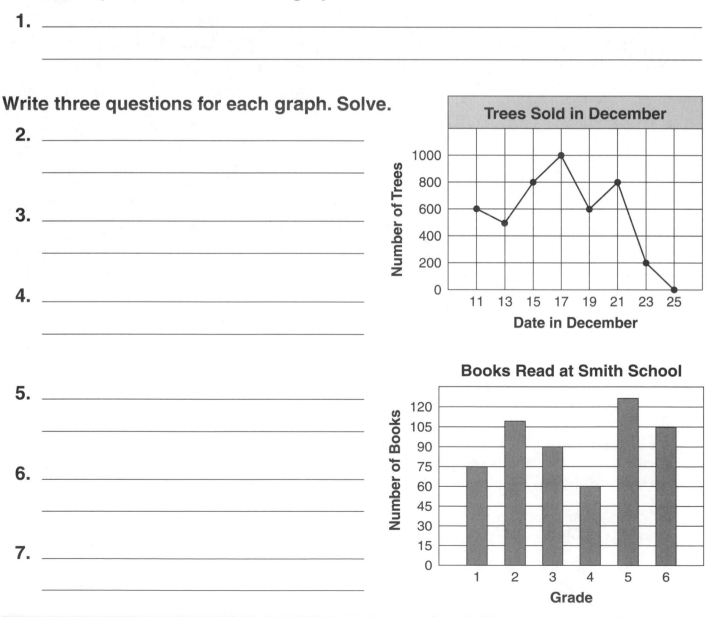

2. _____

_____

3. _____

_____

4. _____

_____

5. _____

_____

6. _____

_____

7. _____

_____

Trees Sold in December

Number of Trees: 1000, 800, 600, 400, 200, 0

Date in December: 11 13 15 17 19 21 23 25

Books Read at Smith School

Number of Books: 120, 105, 90, 75, 60, 45, 30, 15, 0

Grade: 1 2 3 4 5 6

Use with Lesson 7-9, text pages 256–257.

Problem-Solving: Review of Strategies

Name _____

Date _____

Solve. Name the strategy.

1. Lalita is planning a club dinner. There will be 25 minutes for speakers. She allows 7 minutes for each speaker and schedules as many as possible. Can she schedule another speaker for 5 minutes?

_____

2. Alex gave 3 baseball cards to his sister. Then he traded half of the cards he had left to his brother for a baseball magazine. He put the remaining 12 cards into an album. How many cards did he start with?

_____

3. Hank has 2 baseball hats and 4 T-shirts. The hats are blue and red. The shirts are red, white, green and blue. How many ways can he wear the hats and shirts together?

_____

4. Charlene bought 3 pictures of horses for $3.85 each. She gave the cashier $15. How much change did she get back? What bills and coins might she receive?

_____

5. Clare, Norma, and Ricardo work in the library after school. Norma does not work the most hours, but she works more hours than Clare. Who works the most hours? the fewest?

_____

6. Jermaine has nickels and dimes worth $1.00. He has 3 times as many nickels as dimes. How many nickels does Jermaine have? how many dimes?

_____

7. Deven and Raj collected seashells at the beach. Devon collected 79 more shells than Raj, who collected 276 shells. How many seashells did Devon collect?

_____

8. On Monday, Lou had 133 stickers and Lana had 185 stickers. By Wednesday, Lou had 158 stickers and Lana had 197. How many stickers did Lana get between Monday and Wednesday?

_____

Writing Fractions

Name _____

Date _____

The **numerator** names the number of equal parts that are dotted. → $\frac{3}{5}$ ← The **denominator** names the total number of equal parts.

or

Read: three fifths

Write: $\frac{3}{5}$

Use the fractions in the box at the right.

1. List the fractions that have a

   a. numerator of 3. _____

   b. denominator of 8. _____

   c. numerator of 5. _____

   d. denominator of 5. _____

$\frac{3}{10}$	$\frac{7}{10}$	$\frac{3}{5}$	$\frac{1}{3}$
$\frac{5}{8}$	$\frac{2}{3}$	$\frac{1}{8}$	
$\frac{1}{5}$	$\frac{4}{5}$		
$\frac{7}{8}$	$\frac{3}{6}$	$\frac{3}{8}$	
	$\frac{5}{6}$		

Write each as a fraction.

2. one half _____

3. six tenths _____

4. three fourths _____

5. two thirds _____

6. four fifths _____

7. four eighths _____

Write each fraction in words.

8. $\frac{2}{10}$ _____

9. $\frac{1}{3}$ _____

10. $\frac{2}{8}$ _____

11. $\frac{4}{6}$ _____

12. $\frac{1}{4}$ _____

13. $\frac{5}{12}$ _____

PROBLEM SOLVING

14. Sharene wrote a fraction with a numerator of 7. Its denominator was more than its numerator. What fraction did she write? _____

Estimating Fractions

Name _____

Date _____

About what fraction of each region is shaded?

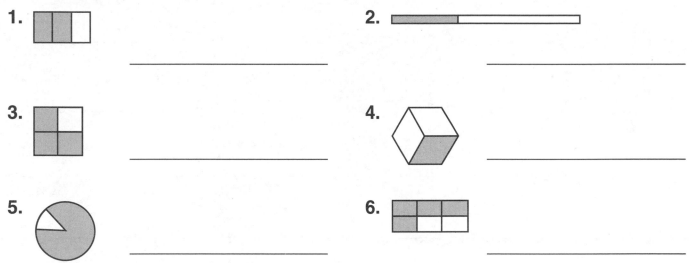

$\frac{1}{2}$ shaded   more than $\frac{1}{2}$ less than $\frac{1}{2}$

Write *more than half* or *less than half*.
Then tell about what fraction of each region is shaded.

1. ⬛ _____

2. ▭ _____

3. ⬛ _____

4. ⬦ _____

5. ◔ _____

6. ⬛ _____

Use the number line. Write whether each fraction is *closer to 0*,
***closer to* $\frac{1}{2}$, or *closer to 1*.**

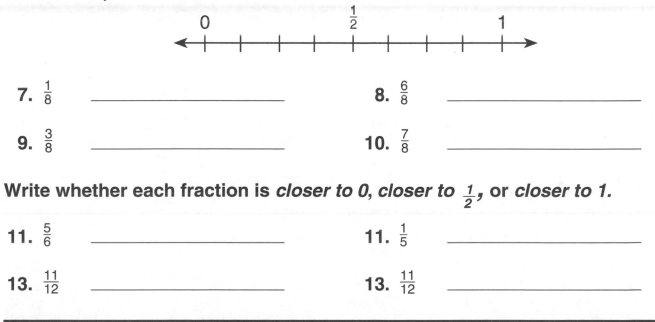

7. $\frac{1}{8}$ _____

8. $\frac{6}{8}$ _____

9. $\frac{3}{8}$ _____

10. $\frac{7}{8}$ _____

Write whether each fraction is *closer to 0*, *closer to* $\frac{1}{2}$, or *closer to 1*.

11. $\frac{5}{6}$ _____

11. $\frac{1}{5}$ _____

13. $\frac{11}{12}$ _____

13. $\frac{11}{12}$ _____

Writing
Equivalent Fractions

Remember: Equivalent fractions name the same part.

$\dfrac{1}{3} = \dfrac{?}{6}$ ← Think: $3 \times 2 = 6$

$\dfrac{1 \times 2}{3 \times 2} = \dfrac{2}{6}$

So $\dfrac{1}{3} = \dfrac{2}{6}$.

$\dfrac{2}{4} = \dfrac{6}{?}$ ← Think: $2 \times 3 = 6$

$\dfrac{2 \times 3}{4 \times 3} = \dfrac{6}{12}$

So $\dfrac{2}{4} = \dfrac{6}{12}$.

Complete.

1. $\dfrac{1 \times 4}{2 \times 4} =$ _____

2. $\dfrac{2 \times 2}{3 \times 2} =$ _____

3. $\dfrac{3 \times 2}{4 \times 2} =$ _____

4. $\dfrac{2 \times}{5 \times 2} =$ _____

5. $\dfrac{1 \times 5}{2 \times} =$ _____

6. $\dfrac{1 \times}{4 \times 3} =$ _____

7. $\dfrac{2 \times}{3 \times} = \dfrac{}{9}$

8. $\dfrac{5 \times}{6 \times} = \dfrac{10}{}$

9. $\dfrac{3 \times}{8 \times} = \dfrac{}{16}$

10. $\dfrac{4}{6} = \dfrac{}{12}$

11. $\dfrac{3}{10} = \dfrac{}{20}$

12. $\dfrac{4}{8} = \dfrac{}{24}$

13. $\dfrac{2}{6} = \dfrac{}{18}$

14. $\dfrac{2}{3} = \dfrac{}{9}$

15. $\dfrac{4}{7} = \dfrac{}{14}$

16. $\dfrac{2}{6} = \dfrac{}{12}$

17. $\dfrac{3}{4} = \dfrac{}{8}$

18. $\dfrac{1}{5} = \dfrac{}{25}$

19. $\dfrac{1}{3} = \dfrac{}{9}$

20. $\dfrac{3}{7} = \dfrac{}{14}$

21. $\dfrac{3}{4} = \dfrac{}{12}$

PROBLEM SOLVING

22. Jamie ate $\frac{1}{4}$ of a pizza. Charles ate $\frac{2}{8}$ of the same pizza. Did they eat the same amount of pizza? Show your work. _____

Use with Lesson 8-5, text pages 274–275.

Factors

> **Common factors** are numbers that are factors of two or more products.
>
> | | |
> |---|---|
> | $1 \times 6 = 6$ | $1 \times 12 = 12$ |
> | $2 \times 3 = 6$ | $2 \times 6 = 12$ |
> | *Factors of 6:* 1, 2, 3, 6 | $3 \times 4 = 12$ |
> | | *Factors of 12:* 1, 2, 3, 4, 6, 12 |
>
> Common factors of 6 and 12: 1, 2, 3, and 6.
> Greatest Common Factor (GCF): 6

Find the missing factor.

1a. $1 \times$ _____ $= 20$ **b.** $2 \times$ _____ $= 20$ **c.** $4 \times$ _____ $= 20$

2a. $1 \times$ _____ $= 18$ **b.** $2 \times$ _____ $= 18$ **c.** $3 \times$ _____ $= 18$

List all the factors of each.

3. 2 _____   **4.** 5 _____

5. 9 _____   **6.** 10 _____

7. 12 _____   **8.** 14 _____

9. 24 _____   **10.** 28 _____

11. 27 _____   **12.** 15 _____

Write all the common factors of each set of numbers.
Then circle the GCF.

13. 8 and 10 _____   **14.** 2 and 10 _____

15. 12 and 15 _____   **16.** 8 and 9 _____

17. 10 and 20 _____   **18.** 6 and 18 _____

19. 4, 12 and 16 _____   **20.** 9, 12, and 18 _____

Fractions: Lowest Terms

Name _____

Date _____

Write in lowest terms: $\frac{6}{8}$

$$\frac{6 \div 2}{8 \div 2} = \frac{3}{4}$$

Factors of 6: 1, 2, 3, 6
Factors of 8: 1, 2, 4, 8
Common factors: 1, 2

Greatest Common Factor: 2

So $\frac{6}{8}$ in lowest terms is $\frac{3}{4}$.

Find the greatest common factor (GCF).

1. 3 and 9

GCF: _____

2. 5 and 10

GCF: _____

3. 6 and 9

GCF: _____

4. 6 and 8

GCF: _____

5. 4 and 12

GCF: _____

6. 3 and 18

GCF: _____

7. 7 and 21

GCF: _____

8. 8 and 12

GCF: _____

Is each fraction in lowest terms? Write *Yes* or *No*.

9. $\frac{6}{9}$ _____

10. $\frac{5}{7}$ _____

11. $\frac{4}{5}$ _____

12. $\frac{9}{12}$ _____

Complete.

13. $\frac{2 \div 2}{4 \div 2} =$ _____

14. $\frac{6 \div 3}{9 \div 3} =$ _____

15. $\frac{4 \div 4}{32 \div 4} =$ _____

16. $\frac{10 \div 5}{15 \div 5} =$ _____

17. $\frac{6 \div 6}{12 \div 6} =$ _____

18. $\frac{9 \div 3}{12 \div 3} =$ _____

19. $\frac{7 \div 7}{28 \div 7} =$ _____

20. $\frac{16 \div 8}{24 \div 8} =$ _____

Write each fraction in lowest terms.

21. $\frac{4}{6} =$ _____

22. $\frac{2}{6} =$ _____

23. $\frac{4}{8} =$ _____

24. $\frac{5}{10} =$ _____

25. $\frac{8}{20} =$ _____

26. $\frac{9}{12} =$ _____

27. $\frac{3}{18} =$ _____

28. $\frac{6}{14} =$ _____

29. $\frac{6}{18} =$ _____

30. $\frac{4}{24} =$ _____

31. $\frac{8}{12} =$ _____

32. $\frac{4}{10} =$ _____

33. $\frac{2}{12} =$ _____

34. $\frac{2}{16} =$ _____

35. $\frac{7}{21} =$ _____

36. $\frac{2}{20} =$ _____

Use with Lesson 8-7, text pages 278–279.

Mixed Numbers

Name _____

Date _____

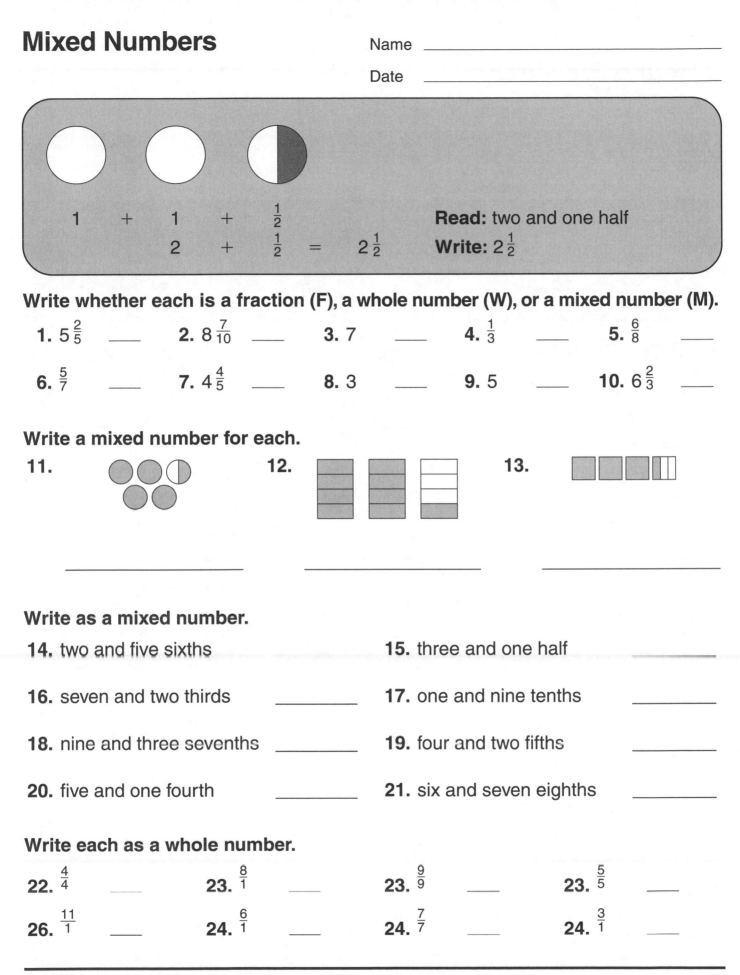

$$1 \quad + \quad 1 \quad + \quad \frac{1}{2}$$
$$2 \quad + \quad \frac{1}{2} \quad = \quad 2\frac{1}{2}$$

Read: two and one half
Write: $2\frac{1}{2}$

Write whether each is a fraction (F), a whole number (W), or a mixed number (M).

1. $5\frac{2}{5}$ ____

2. $8\frac{7}{10}$ ____

3. 7 ____

4. $\frac{1}{3}$ ____

5. $\frac{6}{8}$ ____

6. $\frac{5}{7}$ ____

7. $4\frac{4}{5}$ ____

8. 3 ____

9. 5 ____

10. $6\frac{2}{3}$ ____

Write a mixed number for each.

11. _____

12. _____

13. _____

Write as a mixed number.

14. two and five sixths

15. three and one half ____

16. seven and two thirds ____

17. one and nine tenths ____

18. nine and three sevenths ____

19. four and two fifths ____

20. five and one fourth ____

21. six and seven eighths ____

Write each as a whole number.

22. $\frac{4}{4}$ ____

23. $\frac{8}{1}$ ____

23. $\frac{9}{9}$ ____

23. $\frac{5}{5}$ ____

26. $\frac{11}{1}$ ____

24. $\frac{6}{1}$ ____

24. $\frac{7}{7}$ ____

24. $\frac{3}{1}$ ____

Comparing and
Ordering Fractions

Name _____

Date _____

Compare: $\frac{4}{5}$ _?_ $\frac{3}{5}$

| The denominators are the same. |

Compare the numerators.

$4 > 3$ So $\frac{4}{5} > \frac{3}{5}$.

Compare: $\frac{2}{5}$ _?_ $\frac{9}{20}$

| The denominators are different. |

Rename as equivalent fractions.
With the same denominators: $\frac{2 \times 4}{5 \times 4} = \frac{8}{20}$

Compare the numerators.

$8 < 9$ → $\frac{8}{20} < \frac{9}{20}$ So $\frac{2}{5} < \frac{9}{20}$.

Compare. Write <, =, or >.

1. $\frac{3}{4}$ _____ $\frac{1}{4}$

2. $\frac{1}{6}$ _____ $\frac{5}{6}$

3. $\frac{5}{8}$ _____ $\frac{7}{8}$

4. $\frac{3}{5}$ _____ $\frac{2}{5}$

5. $\frac{7}{8}$ _____ $\frac{5}{8}$

6. $\frac{1}{3}$ _____ $\frac{2}{3}$

7. $\frac{4}{5}$ _____ $\frac{2}{10}$

8. $\frac{2}{5}$ _____ $\frac{4}{10}$

9. $\frac{7}{8}$ _____ $\frac{15}{16}$

10. $\frac{1}{5}$ _____ $\frac{3}{15}$

11. $\frac{1}{5}$ _____ $\frac{1}{10}$

12. $\frac{1}{3}$ _____ $\frac{3}{12}$

13. $\frac{1}{3}$ _____ $\frac{5}{6}$

14. $\frac{7}{8}$ _____ $\frac{1}{4}$

15. $\frac{3}{4}$ _____ $\frac{9}{12}$

16. $\frac{5}{8}$ _____ $\frac{1}{2}$

17. $\frac{9}{10}$ _____ $\frac{4}{5}$

18. $\frac{3}{4}$ _____ $\frac{5}{8}$

19. $2\frac{5}{6}$ _____ $2\frac{1}{6}$

20. $2\frac{1}{8}$ _____ $3\frac{1}{8}$

21. $2\frac{1}{2}$ _____ $2\frac{1}{2}$

22. $4\frac{2}{5}$ _____ $4\frac{2}{5}$

23. $8\frac{1}{2}$ _____ $6\frac{1}{2}$

24. $5\frac{1}{8}$ _____ $5\frac{5}{8}$

PROBLEM SOLVING

25. Of the dogs in the show, $\frac{1}{4}$ were
retrievers and $\frac{3}{8}$ were terriers.
Were there more retrievers or terriers?   _____

26. Two thirds of the dogs were brown and
$\frac{1}{6}$ were black. Were more dogs brown
or black?   _____

Ordering Fractions

Name _____

Date _____

Order from greatest to least: $\frac{3}{10}$, $\frac{1}{5}$, $\frac{9}{10}$

Rename as equivalent fractions
with the same denominator: $\frac{1}{5} = \frac{1 \times 2}{5 \times 2} = \frac{2}{10}$

Compare the numerators: $\frac{3}{10} > \frac{2}{10}$ $\qquad$ $\frac{9}{10} > \frac{3}{10}$

The order from greatest to least: $\frac{9}{10}$, $\frac{3}{10}$, $\frac{1}{5}$

Write in order from least to greatest.

1. $\frac{3}{5}$, $\frac{1}{5}$, $\frac{5}{5}$ _____

2. $\frac{2}{9}$, $\frac{8}{9}$, $\frac{4}{9}$ _____

3. $\frac{5}{8}$, $\frac{7}{8}$, $\frac{3}{8}$ _____

4. $\frac{1}{6}$, $\frac{5}{6}$, $\frac{2}{6}$ _____

5. $\frac{1}{12}$, $\frac{1}{3}$, $\frac{11}{12}$ _____

6. $\frac{1}{5}$, $\frac{9}{10}$, $\frac{7}{10}$ _____

7. $\frac{1}{2}$, $\frac{1}{4}$, $\frac{3}{4}$ _____

8. $\frac{4}{10}$, $\frac{7}{10}$, $\frac{1}{2}$ _____

Write in order from greatest to least.

9. $\frac{2}{6}$, $\frac{4}{6}$, $\frac{3}{6}$ _____

10. $\frac{3}{7}$, $\frac{6}{7}$, $\frac{2}{7}$ _____

11. $\frac{7}{12}$, $\frac{2}{12}$, $\frac{3}{4}$ _____

12. $\frac{2}{5}$, $\frac{3}{15}$, $\frac{4}{15}$ _____

13. $\frac{1}{2}$, $\frac{3}{8}$, $\frac{6}{8}$ _____

14. $\frac{7}{12}$, $\frac{2}{3}$, $\frac{11}{12}$ _____

15. $\frac{1}{2}$, $\frac{4}{6}$, $\frac{5}{6}$ _____

16. $\frac{2}{5}$, $\frac{7}{10}$, $\frac{6}{10}$ _____

PROBLEM SOLVING

17. Dana bought $\frac{3}{4}$ yd, $\frac{7}{8}$ yd, and $\frac{3}{8}$ yd
of cloth. Which was longest? Which
was shortest? _____

Use with Lesson 8-10, text pages 284–285.

Problem-Solving Strategy: Logic and Analogies

Name _____

Date _____

One ribbon is $3\frac{1}{3}$ in. long, another is $3\frac{1}{6}$ in. long, and a third is $3\frac{5}{6}$ in. long. The blue ribbon is longer than the white ribbon, and the white ribbon is longer than the green ribbon. How long is the green ribbon?

Think: $3 = 3 = 3$ The whole numbers for each length are equal. Compare the fractions.

$\frac{1}{3} = \frac{1 \times 2}{3 \times 2} = \frac{2}{6}$ $\frac{5}{6} > \frac{2}{6}$ and $\frac{2}{6} > \frac{1}{6}$

So $3\frac{5}{6}$ is longer than $3\frac{1}{3}$ and $3\frac{1}{3}$ is longer than $3\frac{1}{6}$.

The green ribbon is $3\frac{1}{6}$ in. long.

Solve. Do your work on a separate sheet of paper.

1. Lavonne is shorter than Stan. Carlos is the shortest. One of them is $4\frac{3}{12}$ ft tall another is $4\frac{3}{4}$ ft tall. and the third is $4\frac{5}{12}$ ft tall. How tall is Stan?

   _____

2. Pete's mother is the oldest in his family. His uncle is older than his aunt. Their ages are $71\frac{1}{2}$, $72\frac{3}{4}$, and $73\frac{1}{4}$ years. How old is his uncle?

   _____

3. Lucia cuts boards to build a dog house. One board is $4\frac{7}{12}$ ft long, another is $4\frac{2}{3}$ ft long, and a third is $4\frac{1}{3}$ ft long. The side board is longer than the front board, and the front board is not the shortest board. How long is the front board?

   _____

4. Warren, Tami, and Omar each walked in the park. One walked $3\frac{3}{10}$ miles, another walked $3\frac{4}{10}$ miles, and the third walked $3\frac{3}{5}$ miles. Omar walked the farthest, and Tami walked less than Warren. How far did Warren walk?

   _____

5. Leland weighed his three cats, a calico, a black-and-white, and a gray. They weighed $8\frac{3}{16}$ lb, $8\frac{5}{8}$ lb, and $8\frac{4}{16}$ lb. The calico weighs more than the black-and-white, and the black-and-white is not the lightest cat. How much does the gray cat weigh?

   _____

6. $\frac{2}{3}$ is to $\frac{2}{12}$ as $\frac{5}{4}$ is to _____ .

7. △ is to ◬ as ▢ is to _____ .

8. *XYZ* is to *EFG* as *ZYX* is to _____ .

9. *ABA* is to *QRQ* as *BAB* is to _____ .

Use with Lesson 8-12, text pages 288–289.

Adding: Like Denominators

> Add the numerators. Write the like denominator.
> Write the sum in lowest terms.
>
> $$\frac{1}{6} + \frac{3}{6} = \frac{4}{6} = \frac{2}{3}$$

Add. Write the sum in lowest terms.

1. $\frac{4}{9} + \frac{2}{9} =$ _____

2. $\frac{1}{8} + \frac{3}{8} =$ _____

3. $\frac{7}{12} + \frac{5}{12} =$ _____

4. $\frac{6}{10}$ $+ \frac{3}{10}$

5. $\frac{1}{4}$ $+ \frac{1}{4}$

6. $\frac{3}{6}$ $+ \frac{1}{6}$

7. $\frac{2}{9}$ $+ \frac{5}{9}$

8. $\frac{3}{6}$ $+ \frac{1}{6}$

9. $\frac{4}{10}$ $+ \frac{3}{10}$

10. $\frac{5}{9}$ $+ \frac{1}{9}$

11. $\frac{3}{12}$ $+ \frac{2}{12}$

12. $\frac{4}{8}$ $+ \frac{2}{8}$

13. $\frac{5}{10}$ $+ \frac{5}{10}$

14. $\frac{1}{3}$ $+ \frac{1}{3}$

15. $\frac{2}{8}$ $+ \frac{3}{8}$

16. $\frac{1}{10}$ $+ \frac{3}{10}$

17. $\frac{2}{4}$ $+ \frac{1}{4}$

18. $\frac{3}{5}$ $+ \frac{1}{5}$

19. $\frac{1}{6}$ $+ \frac{2}{6}$

20. $\frac{1}{2}$ $+ \frac{1}{2}$

21. $\frac{3}{8}$ $+ \frac{3}{8}$

22. $\frac{5}{10}$ $+ \frac{3}{10}$

23. $\frac{2}{12}$ $+ \frac{7}{12}$

PROBLEM SOLVING Write each answer in simplest form.

24. Tom used $\frac{1}{4}$ c milk and $\frac{1}{4}$ c water in a recipe. How much liquid did he use in all?

_____

25. What is the sum of one sixth and one sixth? _____

Subtracting: Like Denominators

Name _____

Date _____

Subtract the numerators. Write the like denominator.
Write the difference in lowest terms.

$$\frac{8}{10} - \frac{6}{10} = \frac{2}{10} = \frac{1}{5}$$

Subtract. Write the difference in lowest terms.

1. $\frac{5}{8} - \frac{2}{8} =$ _____

2. $\frac{9}{10} - \frac{2}{10} =$ _____

3. $\frac{11}{12} - \frac{5}{12} =$ _____

4. $\frac{3}{4}$
$- \frac{1}{4}$

5. $\frac{10}{12}$
$- \frac{5}{12}$

6. $\frac{7}{9}$
$- \frac{2}{9}$

7. $\frac{7}{8}$
$- \frac{5}{8}$

8. $\frac{5}{9}$
$- \frac{3}{9}$

9. $\frac{9}{12}$
$- \frac{5}{12}$

10. $\frac{6}{8}$
$- \frac{1}{8}$

11. $\frac{8}{10}$
$- \frac{2}{10}$

12. $\frac{5}{6}$
$- \frac{3}{6}$

13. $\frac{11}{12}$
$- \frac{2}{12}$

14. $\frac{1}{2}$
$- \frac{1}{2}$

15. $\frac{4}{6}$
$- \frac{1}{6}$

16. $\frac{3}{5}$
$- \frac{1}{5}$

17. $\frac{9}{10}$
$- \frac{5}{10}$

18. $\frac{7}{8}$
$- \frac{3}{8}$

19. $\frac{3}{4}$
$- \frac{2}{4}$

20. $\frac{2}{3}$
$- \frac{2}{3}$

21. $\frac{3}{6}$
$- \frac{1}{6}$

22. $\frac{10}{12}$
$- \frac{1}{12}$

23. $\frac{7}{10}$
$- \frac{3}{10}$

PROBLEM SOLVING Write the difference in simplest form.

24. Annie used $\frac{3}{8}$ lb flour for her bread
recipe. Atsuo used $\frac{7}{8}$ lb flour for his
bread recipe. Who used more flour?
How much more?

_____

98 **Use with Lessons 9-2, text pages 298–299.**

Improper Fractions

Write $\frac{16}{3}$ as a mixed number.

Divide the numerator by the denominator.	**Write the quotient as a whole number.**	**Write the remainder over the divisor.**
$\frac{16}{3} = 3)\overline{16}^{\,5\,R1}$	5	$\frac{1}{3}$

So $\frac{16}{3} = 5\frac{1}{3}$.

Write as a whole number or a mixed number in simplest form.

1. $\frac{27}{3}$

2. $\frac{34}{7}$

3. $\frac{56}{9}$

4. $\frac{15}{6}$

_____

5. $\frac{30}{5}$

6. $\frac{37}{8}$

7. $\frac{19}{9}$

8. $\frac{34}{2}$

_____

9. $\frac{25}{6}$

10. $\frac{18}{4}$

11. $\frac{8}{3}$

12. $\frac{11}{5}$

_____

Add or subtract. Write each sum or difference as a mixed number in simplest form.

13. $\frac{3}{7} + \frac{8}{7}$

14. $\frac{30}{8} - \frac{20}{8}$

15. $\frac{14}{9} - \frac{3}{9}$

16. $\frac{5}{6} + \frac{11}{6}$

_____

17. $\frac{16}{8} + \frac{4}{8}$

18. $\frac{25}{4} + \frac{25}{4}$

19. $\frac{85}{5} - \frac{55}{5}$

20. $\frac{13}{3} + \frac{27}{3}$

_____

PROBLEM SOLVING

21. What is sixteen fourths as a whole or mixed number?

23. What is nineteen fifths as a whole or mixed number?

_____

Estimating With Mixed Numbers

Name _____

Date _____

| Estimate.
Add the whole numbers.

$3\frac{1}{3} + 4\frac{1}{6} + 6\frac{2}{3} = \underline{\ ?\ }$
$3\ \ \ + 4\ \ \ + 6\ \ \ = 13$ | Estimate.
Subtract the whole numbers.

$73\frac{2}{8} - 11\frac{7}{8} = \underline{\ ?\ }$
$73\ \ \ - 11\ \ \ = 62$ |

Estimate the sum or the difference. Watch the signs.

1. $24\frac{3}{8}$
$\ \ 9\frac{2}{4}$
$+\ 31\frac{1}{2}$

2. $32\frac{1}{6}$
$\ 11\frac{2}{3}$
$+\ 21\frac{2}{9}$

3. $16\frac{1}{4}$
$\ 23\frac{1}{2}$
$+\ 10\frac{3}{12}$

4. $19\frac{1}{9}$
$\ \ 6\frac{1}{3}$
$+\ 11\frac{4}{9}$

5. $27\frac{4}{6}$
$-\ 23\frac{4}{12}$

6. $42\frac{2}{3}$
$-\ 20\frac{1}{6}$

7. $31\frac{1}{12}$
$-\ \ 6\frac{2}{3}$

8. $16\frac{7}{9}$
$-\ 10\frac{4}{9}$

9. $34\frac{1}{8}$
$\ 10\frac{3}{4}$
$+\ \ 5\frac{5}{8}$

10. $11\frac{2}{3}$
$\ 12\frac{1}{6}$
$+\ 45\frac{3}{12}$

11. $41\frac{2}{3}$
$\ \ 3\frac{3}{6}$
$+\ 16\frac{8}{9}$

12. $49\frac{7}{8}$
$\ 11\frac{1}{2}$
$+\ 31\frac{3}{8}$

13. $22\frac{9}{10}$
$-\ 20\frac{1}{10}$

14. $98\frac{2}{3}$
$-\ 12\frac{5}{6}$

15. $67\frac{1}{2}$
$-\ 34\frac{4}{6}$

16. $50\frac{4}{8}$
$-\ 20\frac{3}{4}$

PROBLEM SOLVING

17. Kim ran $3\frac{1}{3}$ kilometers one day, $4\frac{1}{3}$ kilometers the next day, and $5\frac{5}{6}$ kilometers on the third day. About how far did she run in all? _____

18. Donna biked $5\frac{1}{3}$ mi to get to her friend's house. Then she took the same path and biked home. About how far did she bike in all? _____

Use with Lesson 9-4, text pages 302–303.

Add and Subtract Mixed Numbers

Name _____

Date _____

Add or subtract. Write the answers in simplest form.

$$1\tfrac{1}{4}$$
$$+\ 9\tfrac{1}{4}$$
$$\overline{\tfrac{2}{4}}$$
$$\longrightarrow$$
$$1\tfrac{1}{4}$$
$$+\ 9\tfrac{1}{4}$$
$$\overline{10\tfrac{2}{4}=10\tfrac{1}{2}}$$

$$6\tfrac{4}{9}$$
$$-\ 4\tfrac{1}{9}$$
$$\overline{\tfrac{3}{9}}$$
$$\longrightarrow$$
$$6\tfrac{4}{9}$$
$$-\ 4\tfrac{1}{9}$$
$$\overline{2\tfrac{3}{9}=2\tfrac{1}{3}}$$

Add. Write the sum in simplest form.

1. $4\tfrac{3}{5}$
$+\ 3\tfrac{1}{5}$

2. $8\tfrac{2}{6}$
$+\ 6\tfrac{3}{6}$

3. $6\tfrac{2}{6}$
$+\ 3\tfrac{2}{6}$

4. $16\tfrac{3}{8}$
$+\ 14\tfrac{2}{8}$

5. $19\tfrac{3}{5}$
$+\ 21\tfrac{1}{5}$

6. $23\tfrac{5}{9}$
$+\ 46\tfrac{2}{9}$

7. $1\tfrac{1}{8}$
$+\ 9\tfrac{6}{8}$

8. $8\tfrac{3}{7}$
$+\ 5\tfrac{2}{7}$

9. $56\tfrac{2}{9}$
$+\ 28\tfrac{4}{9}$

10. $97\tfrac{3}{10}$
$+\ 15\tfrac{1}{10}$

Subtract. Write the difference in simplest form.

11. $4\tfrac{4}{5}$
$-\ 3\tfrac{1}{5}$

12. $4\tfrac{9}{10}$
$-\ 2\tfrac{2}{10}$

13. $43\tfrac{3}{4}$
$-\ 28\tfrac{1}{4}$

14. $80\tfrac{5}{7}$
$-\ 46\tfrac{3}{7}$

15. $17\tfrac{6}{12}$
$-\ 5\tfrac{2}{12}$

16. $56\tfrac{5}{6}$
$-\ 49\tfrac{1}{6}$

17. $98\tfrac{7}{8}$
$-\ 27\tfrac{3}{8}$

18. $65\tfrac{2}{3}$
$-\ 37\tfrac{1}{3}$

19. $86\tfrac{8}{10}$
$-\ 55\tfrac{4}{10}$

20. $23\tfrac{3}{10}$
$-\ 14\tfrac{1}{10}$

Add or subtract. Watch the signs.

21. $8\tfrac{3}{6}$
$+\ 3\tfrac{1}{6}$

22. $64\tfrac{3}{5}$
$+\ 49\tfrac{1}{5}$

23. $87\tfrac{6}{8}$
$-\ 19\tfrac{4}{8}$

24. $17\tfrac{3}{9}$
$+\ 38\tfrac{1}{9}$

25. $8\tfrac{3}{12}$
$-\ 5\tfrac{1}{12}$

PROBLEM SOLVING

26. What is the sum of $9\tfrac{5}{8}$ and $3\tfrac{1}{8}$?
What is the difference? _____

Multiples

Name _____

Date _____

> Find the **least common multiple (LCM)** of 5 and 8.
> Write the multiples.
>
>   Multiples of 5: 0, 5, 10, 15, 20, 25, 30, 35, **40**, . . .
>   Multiples of 8: 0, 8, 16, 24, 32, **40**, . . .
>
> The LCM of 5 and 8 is 40.

Write the first ten multiples of each.

1. 7   ____ ____ ____ ____ ____ ____ ____ ____ ____ ____

2. 3   ____ ____ ____ ____ ____ ____ ____ ____ ____ ____

3. 10   ____ ____ ____ ____ ____ ____ ____ ____ ____ ____

4. 6   ____ ____ ____ ____ ____ ____ ____ ____ ____ ____

Write the first ten multiples. Then write the first two common multiples. Identify the least common multiple (LCM).

5. 2, 5   _____

                  _____

                  _____, _____   LCM _____

6. 8, 2   _____

                  _____

                  _____, _____   LCM _____

7. 3, 6, 9   _____

                  _____

                  _____

                  _____, _____   LCM _____

Write the LCM for each set of numbers.

8. 3, 5, 6 _____   **9.** 2, 4, 5 _____

10. 4, 6, 9 _____   **11.** 4, 6, 8 _____

12. 3, 4, 6 _____   **13.** 4, 5, 8 _____

14. 2, 3, 6 _____   **15.** 2, 3, 5 _____

Adding: Unlike Denominators

Name _____

Date _____

$$\frac{3}{8} = \frac{3}{8}$$
$$+ \frac{1}{4} = \frac{1 \times 2}{4 \times 2} = + \frac{2}{8}$$
$$\frac{5}{8}$$

$$\frac{7}{10} = \frac{7}{10}$$
$$+ \frac{1}{2} = \frac{1 \times 5}{2 \times 5} = + \frac{5}{10}$$
$$\frac{12}{10} = 1\frac{2}{10} = 1\frac{1}{5}$$

Add. Write the sum in simplest form.

1. $\frac{5}{8} =$
$+ \frac{1}{4} =$

2. $\frac{2}{3} =$
$+ \frac{1}{9} =$

3. $\frac{4}{5} =$
$+ \frac{2}{10} =$

4. $\frac{5}{8} =$
$+ \frac{3}{4} =$

5. $\frac{4}{9} =$
$+ \frac{1}{3} =$

6. $\frac{5}{6} =$
$+ \frac{2}{3} =$

7. $\frac{3}{5} =$
$+ \frac{4}{10} =$

8. $\frac{2}{3} =$
$+ \frac{2}{6} =$

9. $\frac{3}{8} =$
$+ \frac{1}{2} =$

10. $\frac{7}{10} =$
$+ \frac{1}{5} =$

11. $\frac{5}{6} =$
$+ \frac{1}{2} =$

12. $\frac{1}{6} =$
$+ \frac{2}{3} =$

13. $\frac{2}{8} =$
$+ \frac{1}{4} =$

14. $\frac{4}{9} =$
$+ \frac{2}{3} =$

15. $\frac{4}{6} =$
$+ \frac{1}{3} =$

16. $\frac{3}{4} =$
$+ \frac{4}{8} =$

PROBLEM SOLVING Write the answer in simplest form.

17. What is the sum of $\frac{5}{6}$ and $\frac{1}{2}$? _____

18. One ninth plus two thirds equal what number? _____

19. Rosa needs $\frac{1}{2}$ yard of material to finish her dress. Alice needs $\frac{5}{8}$ yard. How much material do they need in all? _____

20. Henry worked $\frac{1}{4}$ hour. Yoshi worked $\frac{1}{2}$ hour. How much time did they work together? _____

Subtracting: Unlike Denominators

Name _____

Date _____

$$\begin{array}{rcl} \frac{5}{8} & = & \frac{5}{8} \\ -\frac{1}{4} = \frac{1\times2}{4\times2} & = & -\frac{2}{8} \\ & & \frac{3}{8} \end{array}$$

$$\begin{array}{rcl} \frac{6}{9} & = & \frac{6}{9} \\ -\frac{1}{3} = \frac{1\times3}{3\times3} & = & -\frac{3}{9} \\ & & \frac{3}{9} = \frac{1}{3} \end{array}$$

Subtract. Write the difference in simplest form.

1. $\frac{5}{8} =$
$-\frac{2}{4} =$

2. $\frac{3}{4} =$
$-\frac{3}{8} =$

3. $\frac{7}{12} =$
$-\frac{2}{6} =$

4. $\frac{7}{9} =$
$-\frac{2}{3} =$

5. $\frac{8}{9} =$
$-\frac{1}{3} =$

6. $\frac{4}{5} =$
$-\frac{3}{10} =$

7. $\frac{3}{5} =$
$-\frac{4}{15} =$

8. $\frac{7}{12} =$
$-\frac{1}{4} =$

9. $\frac{6}{10} =$
$-\frac{3}{5} =$

10. $\frac{5}{9} =$
$-\frac{1}{3} =$

11. $\frac{6}{8} =$
$-\frac{2}{4} =$

12. $\frac{5}{8} =$
$-\frac{2}{16} =$

13. $\frac{9}{10} =$
$-\frac{2}{5} =$

14. $\frac{5}{6} =$
$-\frac{2}{3} =$

15. $\frac{7}{8} =$
$-\frac{1}{4} =$

16. $\frac{2}{3} =$
$-\frac{8}{24} =$

PROBLEM SOLVING **Write each answer in simplest form.**

17. What is the difference between $\frac{7}{10}$ and $\frac{2}{5}$? _____

18. $\frac{3}{4}$ minus $\frac{2}{8}$ equals what number? _____

19. A jar contained $\frac{8}{16}$ lb of jelly. Marie used $\frac{1}{8}$ lb. How much jelly is left in the jar? _____

20. What is the difference between $\frac{1}{2}$ and $\frac{9}{10}$? _____

Use with Lesson 9-8, text pages 310–311.

Computing Probability

Name _____

Date _____

> There are 1 apple, 2 nectarines, and 3 pears in a bowl.
> What is the probability of picking a nectarine without looking?
>
> The probability is 2 out of 6.
>
> Write: P (nectarines) $= \frac{2}{6}$

Find the probability of each event. Pick the coins without looking.

1. P (penny) = _____

2. P (nickel) = _____

3. P (dime) = _____

4. P (dime or nickel) = _____

5. P (penny or nickel) = _____

6. P (penny or dime) = _____

Find the probability of each event. Pick the glasses without looking.

7. P (orange) = _____

8. P (tomato) = _____

9. P (grape) = _____

10. P (grapefruit) = _____

11. P (orange or grape) = _____

12. P (grape or tomato) = _____

orange juice

tomato juice

grape juice

Find the probability of each event.

There are 8 marbles in a box. There are 3 red marbles,
2 black marbles and 3 white marbles. Without looking,
what is:

13. P (red) = _____

14. P (red or white) = _____

15. P (black or white) = _____

16. P (not black) = _____

17. P (red or black or white) = _____

Finding Parts of Numbers

Name _____

Date _____

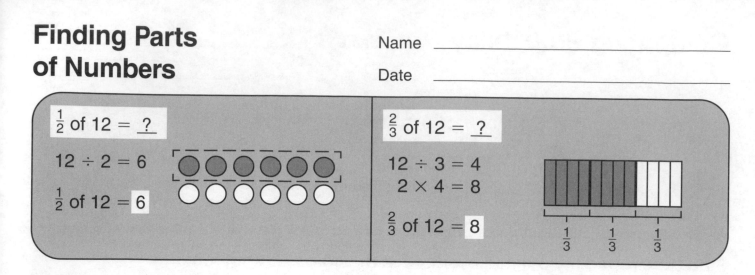

$\frac{1}{2}$ of 12 = _?_

12 ÷ 2 = 6

$\frac{1}{2}$ of 12 = **6**

$\frac{2}{3}$ of 12 = _?_

12 ÷ 3 = 4

2 × 4 = 8

$\frac{2}{3}$ of 12 = **8**

$\frac{1}{3}$ $\frac{1}{3}$ $\frac{1}{3}$

Color to show the part of each number.

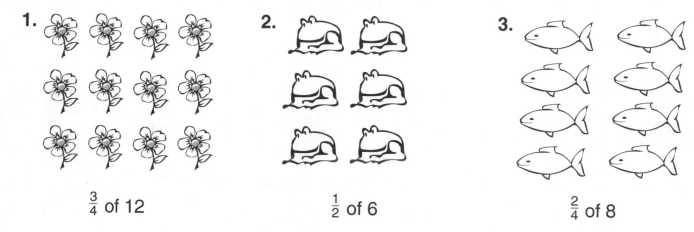

1.

$\frac{3}{4}$ of 12

2.

$\frac{1}{2}$ of 6

3.

$\frac{2}{4}$ of 8

Find the missing number.

4. $\frac{1}{8}$ of 16 = _____

5. $\frac{1}{6}$ of 24 = _____

6. $\frac{1}{5}$ of 30 = _____

7. $\frac{1}{4}$ of 20 = _____

8. $\frac{1}{2}$ of 14 = _____

9. $\frac{1}{4}$ of 36 = _____

10. $\frac{3}{4}$ of 24 = _____

11. $\frac{2}{3}$ of 18 = _____

12. $\frac{3}{5}$ of 45 = _____

13. $\frac{7}{8}$ of 32 = _____

14. $\frac{5}{8}$ of 40 = _____

15. $\frac{3}{8}$ of 48 = _____

16. $\frac{8}{9}$ of 72 = _____

17. $\frac{5}{6}$ of 54 = _____

18. $\frac{4}{7}$ of 28 = _____

PROBLEM SOLVING

19. Ms. Murphy has 36 papers to check. She checked $\frac{3}{4}$ of them. How many did she check? How many more has she left to do?

_____

Problem-Solving Strategy: Use Simpler Numbers

Name _____

Date _____

Sonu wanted to run a total of 6 miles. She ran $2\frac{1}{10}$ miles on Monday, $3\frac{1}{10}$ miles on Tuesday, and $1\frac{3}{10}$ miles on Wednesday. Did she run 6 miles?

Use simpler numbers like 2, 3, and 1.

Add the whole numbers: $2 + 3 + 1 = 6$.

Now compute.

$$\begin{array}{r} 2\frac{1}{10} \\ 3\frac{1}{10} \\ + 1\frac{3}{10} \\ \hline 6\frac{5}{10} = 6\frac{1}{2} \end{array}$$

Sonu ran more than 6 miles.

Solve using simpler numbers. Do your work on a separate sheet of paper.

1. John fenced off the horse pen. The three sides measured $25\frac{1}{6}$ ft, $36\frac{2}{6}$ ft, and $40\frac{2}{6}$ ft. How much fencing did John use?

2. A horse drinks $4\frac{7}{8}$ gallons of water a day. A pony drinks $3\frac{1}{8}$ gallons. How much more does the horse drink than the pony?

3. There are 8 quarts of grain in the bin. If two horses together eat $2\frac{1}{4}$ quarts of grain a day, can they eat for three days?

4. Beth needed to make two lead lines for the horses. Her length of rope was $15\frac{1}{2}$ feet. Did she have enough rope to make two $7\frac{1}{4}$-foot lead lines?

5. Horses ate $30\frac{1}{4}$ bales of hay in December, $28\frac{1}{4}$ bales in January, and $32\frac{1}{4}$ bales in February. How much hay did they eat during those three months?

6. A bread recipe calls for $5\frac{1}{2}$ cups of white flour and $1\frac{1}{2}$ cups of rye flour. How much more white flour than rye flour is needed?

Points, Lines, and Line Segments

Name _____

Date _____

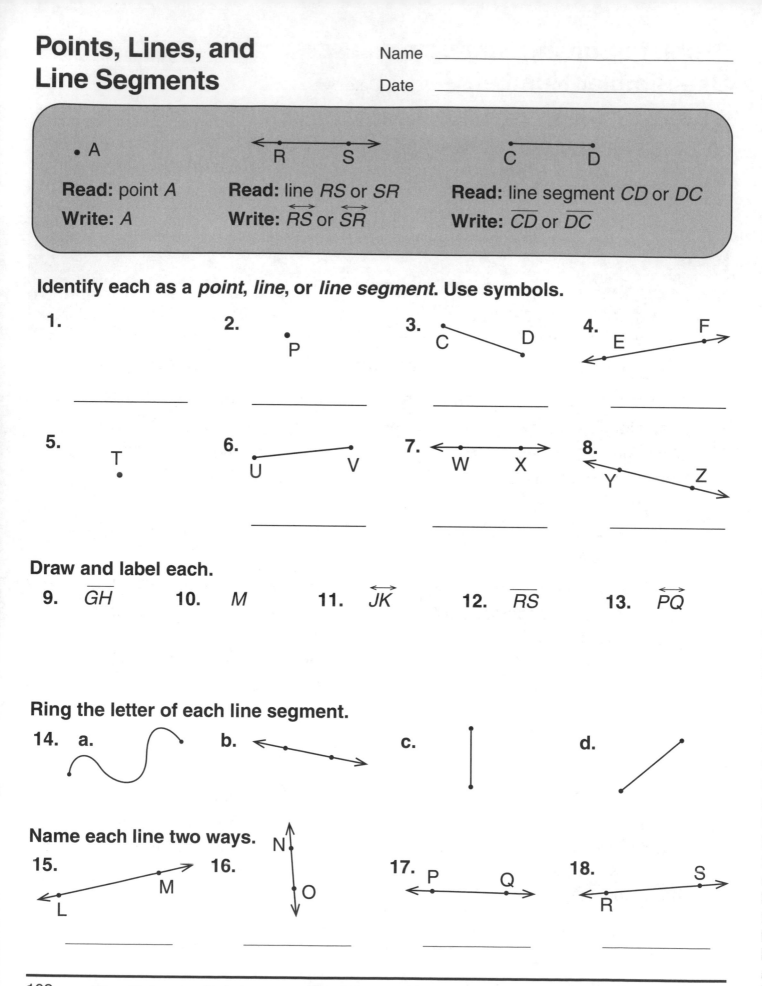

• A	

Read: point A
Write: A

Read: line *RS* or *SR*
Write: $\overleftrightarrow{RS}$ or $\overleftrightarrow{SR}$

Read: line segment *CD* or *DC*
Write: $\overline{CD}$ or $\overline{DC}$

Identify each as a *point*, *line*, or *line segment*. Use symbols.

1. _____

2. •P _____

3. C — D _____

4. E — F _____

5. T • _____

6. U — V _____

7. W — X _____

8. Y — Z _____

Draw and label each.

9. $\overline{GH}$ 10. M   11. $\overleftrightarrow{JK}$ 12. $\overline{RS}$ 13. $\overleftrightarrow{PQ}$

Ring the letter of each line segment.

14. a.   b.   c.   d.

Name each line two ways.

15. L — M _____

16. N — O _____

17. P — Q _____

18. R — S _____

Rays and Angles

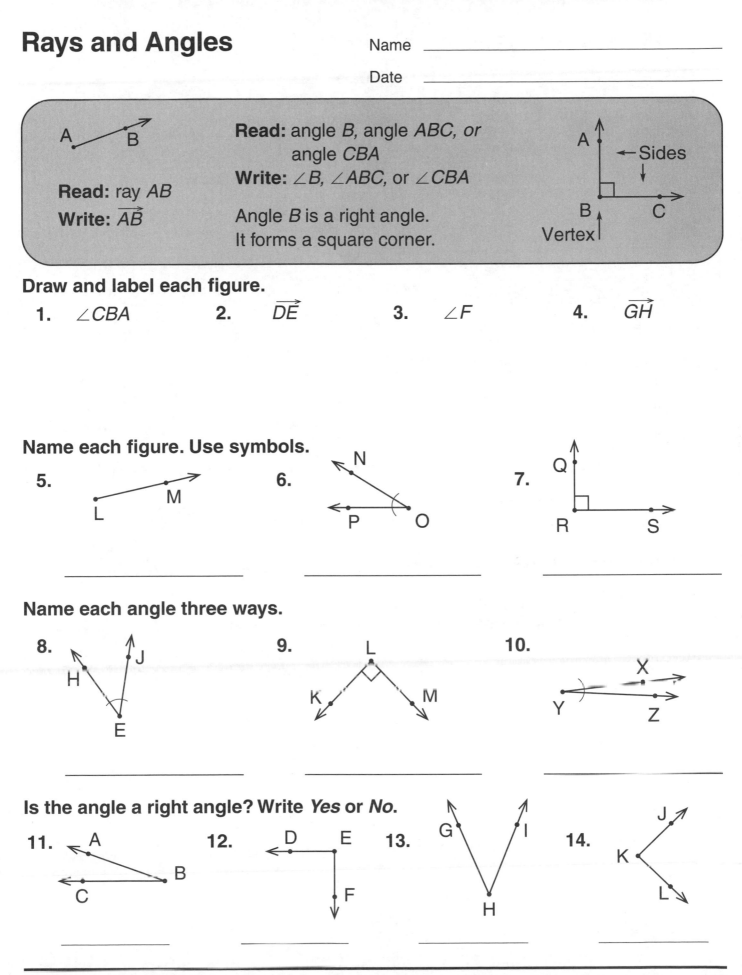

Read: ray *AB*

Write: $\overrightarrow{AB}$

Read: angle *B*, angle *ABC, or* angle *CBA*

Write: ∠*B*, ∠*ABC,* or ∠*CBA*

Angle *B* is a right angle. It forms a square corner.

← Sides

Vertex

Draw and label each figure.

1. ∠*CBA*　　　**2.** $\overrightarrow{DE}$　　　**3.** ∠*F*　　　**4.** $\overrightarrow{GH}$

Name each figure. Use symbols.

5.　　　　　**6.**　　　　　**7.**

_____　　　_____　　　_____

Name each angle three ways.

8.　　　　　**9.**　　　　　**10.**

_____　　　_____　　　_____

Is the angle a right angle? Write *Yes* or *No.*

11.　　　**12.**　　　**13.**　　　**14.**

_____　　_____　　_____　　_____

Parallel and Perpendicular Lines

Name _____

Date _____

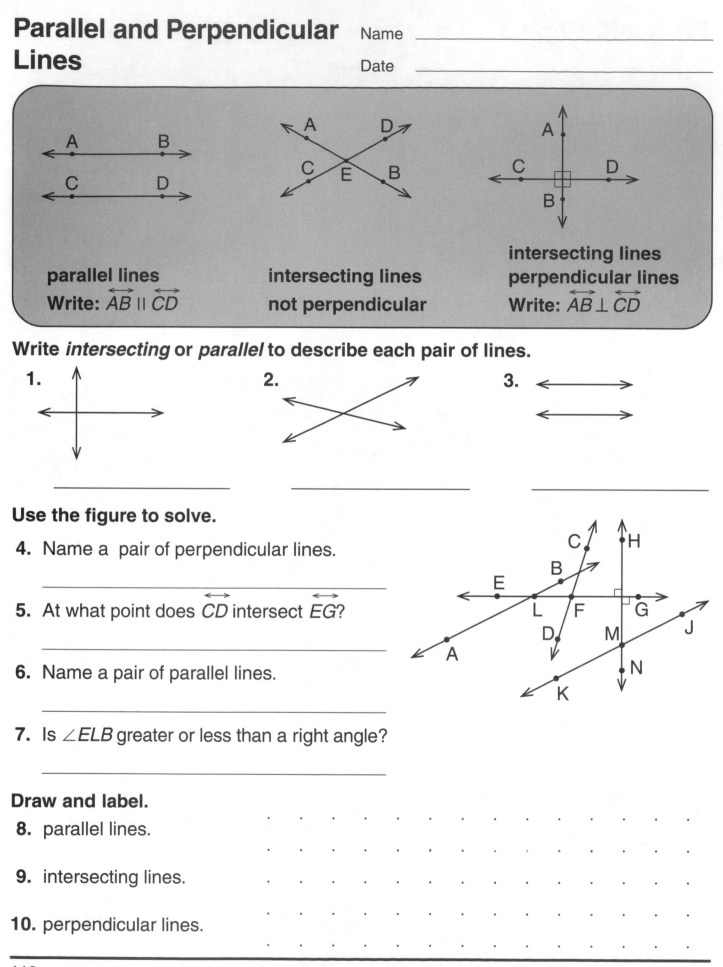

parallel lines
Write: $\overleftrightarrow{AB} \parallel \overleftrightarrow{CD}$

intersecting lines
not perpendicular

intersecting lines
perpendicular lines
Write: $\overleftrightarrow{AB} \perp \overleftrightarrow{CD}$

Write *intersecting* or *parallel* to describe each pair of lines.

1.

_____

2.

_____

3.

_____

Use the figure to solve.

4. Name a pair of perpendicular lines.

_____

5. At what point does $\overleftrightarrow{CD}$ intersect $\overleftrightarrow{EG}$?

_____

6. Name a pair of parallel lines.

_____

7. Is $\angle ELB$ greater or less than a right angle?

_____

Draw and label.

8. parallel lines.

9. intersecting lines.

10. perpendicular lines.

Circles

Name _____

Date _____

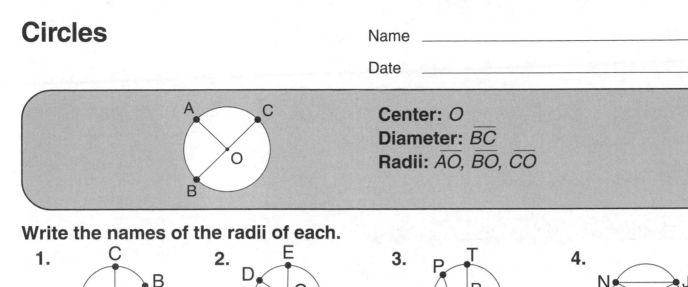

Center: *O*
Diameter: $\overline{BC}$
Radii: $\overline{AO}$, $\overline{BO}$, $\overline{CO}$

Write the names of the radii of each.

1.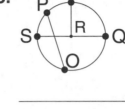

_____

2.

_____

3.

_____

4.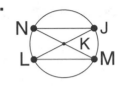

_____

Write the name of each diameter.

5.

_____

6.

_____

7.

8.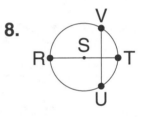

Solve. Use the circle at the right.

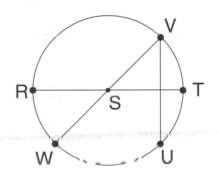

9. Name three line segments that are radii.

_____

10. Name all the diameters shown.

_____

11. Is $\overline{ST}$ a radius? Explain why or why not.

12. Name the circle and its center.

_____

13. If you know $\overline{ST}$ is 6 inches long, can you tell how long $\overline{VU}$ is? What about $\overline{VW}$? Explain your answers.

14. Draw a radius of circle *R*.

_____

Polygons

Name _____

Date _____

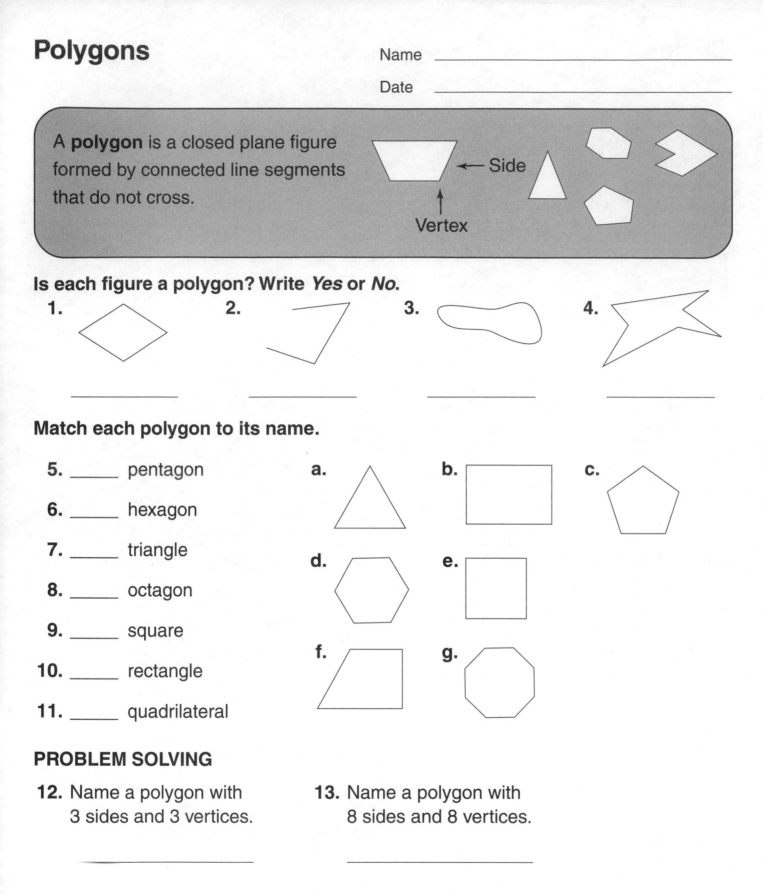

A **polygon** is a closed plane figure formed by connected line segments that do not cross.

← Side

↑ Vertex

Is each figure a polygon? Write *Yes* or *No*.

1. _____

2. _____

3. _____

4. _____

Match each polygon to its name.

5. _____ pentagon

6. _____ hexagon

7. _____ triangle

8. _____ octagon

9. _____ square

10. _____ rectangle

11. _____ quadrilateral

a.

b.

c.

d.

e.

f.

g.

PROBLEM SOLVING

12. Name a polygon with 3 sides and 3 vertices.

_____

13. Name a polygon with 8 sides and 8 vertices.

_____

14. Draw a polygon that has 4 sides and is *not* a square or rectangle.

Use with Lesson 10-5, text pages 334–335.

Quadrilaterals

Name _____

Date _____

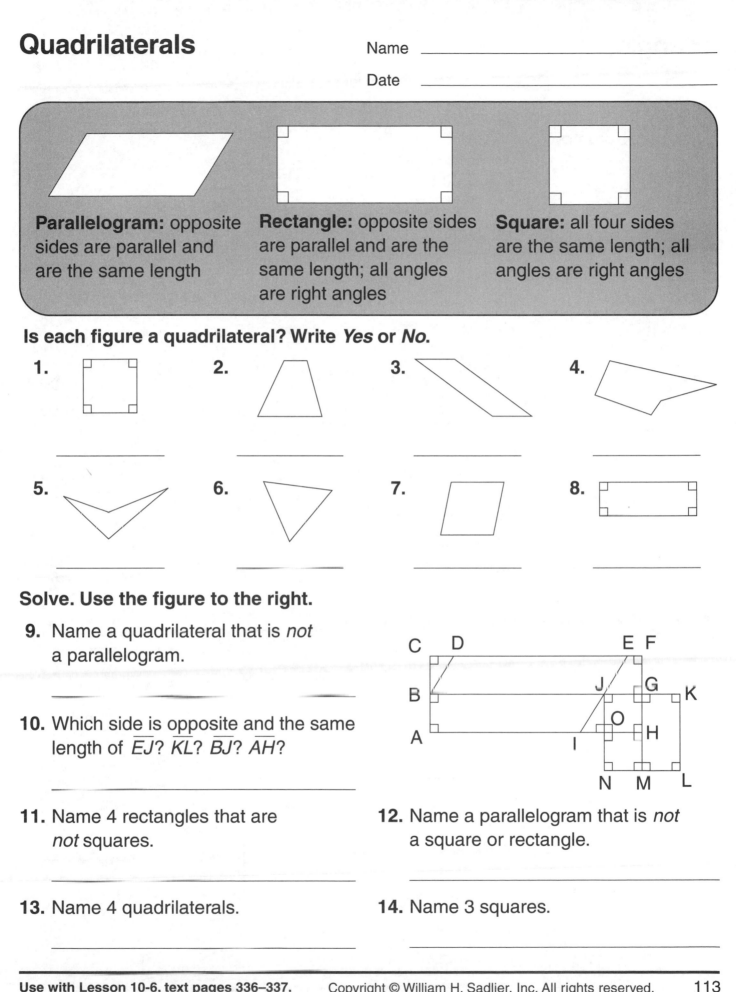

Parallelogram: opposite sides are parallel and are the same length

Rectangle: opposite sides are parallel and are the same length; all angles are right angles

Square: all four sides are the same length; all angles are right angles

Is each figure a quadrilateral? Write *Yes* or *No*.

1. _____

2. _____

3. _____

4. _____

5. _____

6. _____

7. _____

8. _____

Solve. Use the figure to the right.

9. Name a quadrilateral that is *not* a parallelogram.

_____

10. Which side is opposite and the same length of $\overline{EJ}$? $\overline{KL}$? $\overline{BJ}$? $\overline{AH}$?

_____

11. Name 4 rectangles that are *not* squares.

12. Name a parallelogram that is *not* a square or rectangle.

_____

13. Name 4 quadrilaterals.

14. Name 3 squares.

_____

Triangles

Right triangles always have a right angle. **Equilateral triangles** have 3 sides that are the same length.

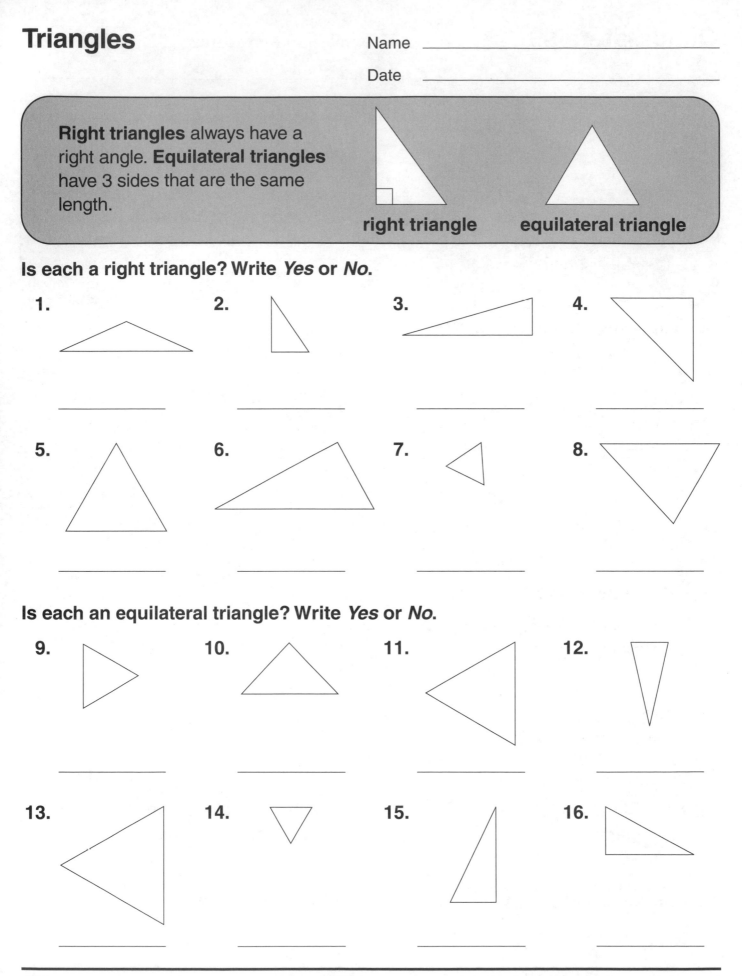

right triangle   **equilateral triangle**

Is each a right triangle? Write *Yes* or *No*.

1. _____

2. _____

3. _____

4. _____

5. _____

6. _____

7. _____

8. _____

Is each an equilateral triangle? Write *Yes* or *No*.

9. _____

10. _____

11. _____

12. _____

13. _____

14. _____

15. _____

16. _____

114   **Use with Lesson 10-7, text pages 338–339.**

Similar Figures

Name _____

Date _____

Similar figures have exactly the same shape. They may or may not be the same size.

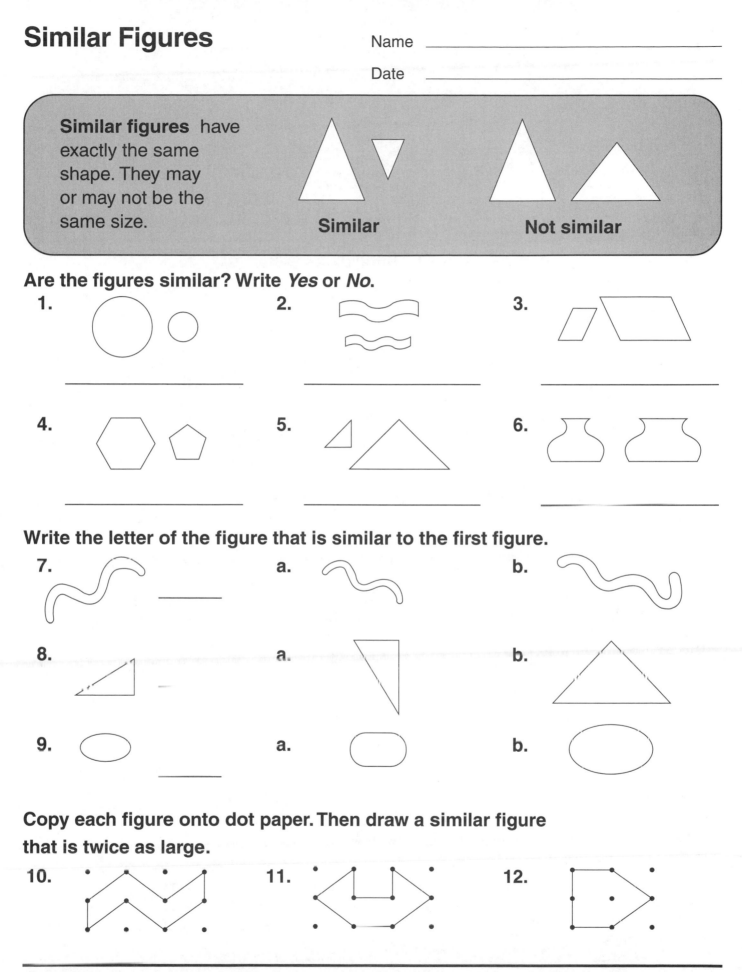

Similar   **Not similar**

Are the figures similar? Write *Yes* or *No*.

1. _____

2. _____

3. _____

4. _____

5. _____

6. _____

Write the letter of the figure that is similar to the first figure.

7. _____   a.   b.

8.   a.   b.

9.   a.   b.

Copy each figure onto dot paper. Then draw a similar figure that is twice as large.

10.   11.   12.

Slides and Flips

Name _____

Date _____

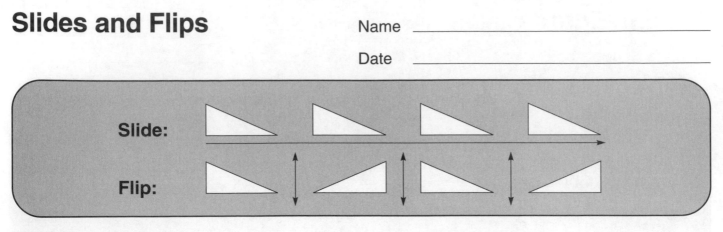

Slide:

Flip:

Draw the next figure in the pattern. Then write *slide* or *flip*
to tell how the pattern was made.

1.

2.

3.

4.

5.

6.

7.

8.

Turns

Turns:

The tracing and the drawing match exactly.
The figure has **half-turn symmetry**.

$\frac{1}{4}$ turn $\frac{1}{2}$ turn

Does each show a turn? Write *Yes* or *No*.

1. _____

2. _____

3. _____

4. _____

5. _____

6. _____

7. _____

8. _____

Does each figure have half-turn symmetry?
Write *Yes* or *No*.

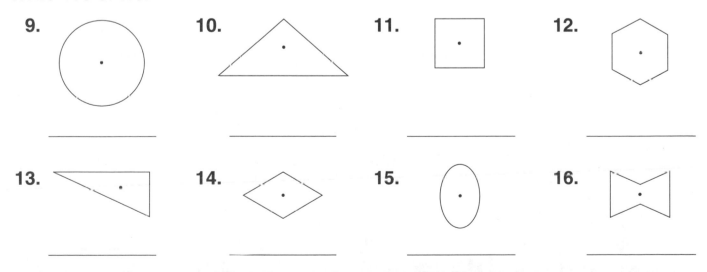

9. _____

10. _____

11. _____

12. _____

13. _____

14. _____

15. _____

16. _____

117

Coordinate Geometry

Name _____

Date _____

In an **ordered pair,** the first number tells you to move to the right. The second number tells you to move up.

Point *M* is located at (4,2).

(0,3) gives the location of point *F*.

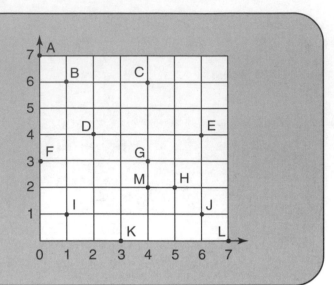

Use the grid above to answer each question.

1. What ordered pair gives the location of point *C*?

   _____

2. What point is located at (5,2)?

   _____

3. What ordered pair gives the location of point *B*?

   _____

4. What point is located at (0,7)?

   _____

5. What point is located at (7,0)?

   _____

6. What ordered pair gives the location of point *E*?

   _____

Use the grid. Write the ordered pairs that will spell the following words.

7. B A G _____

8. B A K E _____

9. C H I L D _____

10. Find another word you can spell using the letters in the grid. Write the ordered pair for each letter in the word. _____

Use with Lesson 10-11, text pages 346–347.

Problem Solving:
Find a Pattern

Name _____

Date _____

Walter draws this pattern of squares. If he continues the pattern, what will be the distance around the 7th square?

5 cm
3 cm
1 cm

Think: Find the pattern.   1, 3, and 5 are consecutive odd numbers.
Extend the pattern to 7 terms.   1, 3, 5, 7, 9, 11, 13
Add to find the distance.   13 + 13 + 13 + 13 = 52

The distance around the 7th square will be 52 cm.

PROBLEM SOLVING Do your work on a separate sheet of paper.

1. George notices that this pattern uses the number of sides that each polygon has. Draw the missing shapes.

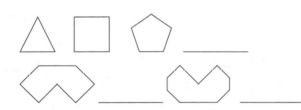

2. A tile border follows this pattern. How many tiles are in a 6 ft border?

2 ft

3. Michelle is painting a pattern along the border of a poster. She has painted a circle, a pentagon, a circle, a parallelogram, a circle, a pentagon, a circle, a parallelogram, a circle, a pentagon and a circle. If she continues the pattern, what shape will come next?

4. Julio stacks six rows of boxes in a pattern. Each box is the same size. There are 13 boxes on the bottom row, 11 boxes on the row above that and 9 boxes in the next row. How many boxes will be in the top row?

5. A design has 2 stars in the first row, 4 stars in the second row, 8 stars in the third row, and 16 stars in the fourth row. If this pattern continues, how many stars will be in the sixth row?

6. Joanna is drawing a pattern. Draw the next shape.

Using Perimeter Formulas

Name _____

Date _____

Formula for the perimeter of a rectangle:

$P = 2 \times \ell \; + \; 2 \times w$
$P = (2 \times 32) + (2 \times 8)$
$P = \quad 64 \quad + \quad 16$
$P = 80$ ft

32 ft

8 ft   8 ft

32 ft

The perimeter is 80 ft.

Find the perimeter of each figure.

1.

19 in.

_____

_____

2.

15 cm

7 cm

_____

_____

3.

143 mm

_____

_____

4.

3 yd

4 yd

_____

_____

5.

25 mi

_____

_____

6.

8 m

_____

_____

PROBLEM SOLVING

7. What is the perimeter of a rectangular vegetable garden with a width of 8 m and a length of 17 m?

_____

8. What is the perimeter of a hexagon, the sides of which all measure 79 cm?

_____

Area

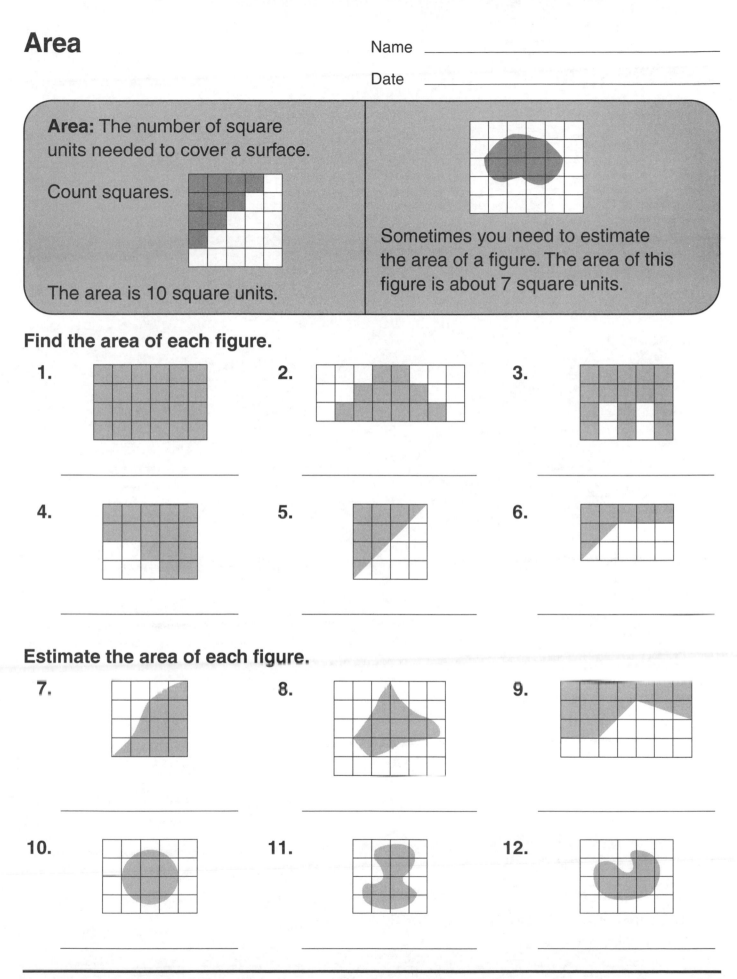

Area: The number of square units needed to cover a surface.

Count squares.

The area is 10 square units.

Sometimes you need to estimate the area of a figure. The area of this figure is about 7 square units.

Find the area of each figure.

1.

2.

3.

4.

5.

6.

Estimate the area of each figure.

7.

8.

9.

10.

11.

12.

Using the Area Formula

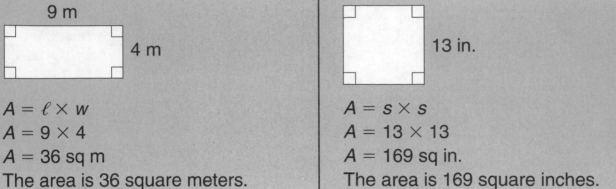

$A = \ell \times w$
$A = 9 \times 4$
$A = 36$ sq m
The area is 36 square meters.

$A = s \times s$
$A = 13 \times 13$
$A = 169$ sq in.
The area is 169 square inches.

Find the area.

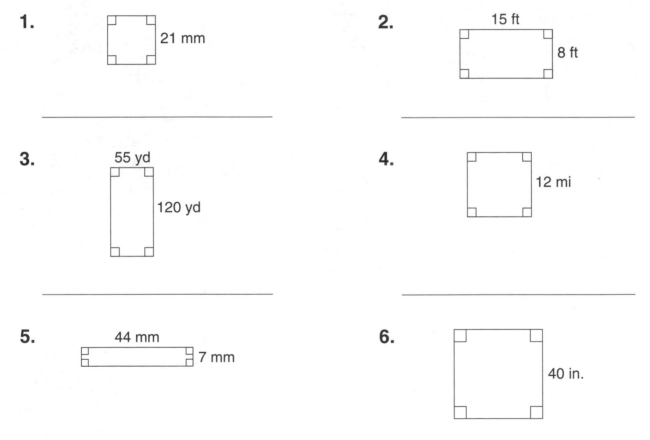

1. 21 mm

_____

2. 15 ft / 8 ft

_____

3. 55 yd / 120 yd

_____

4. 12 mi

_____

5. 44 mm / 7 mm

_____

6. 40 in.

_____

PROBLEM SOLVING

7. Arnie's pool cover is a rectangle that
is 15 meters long and 12 meters wide.
What is its area?

_____

Use with Lesson 11-3, text pages 362–363.

Space Figures

Name _____

Date _____

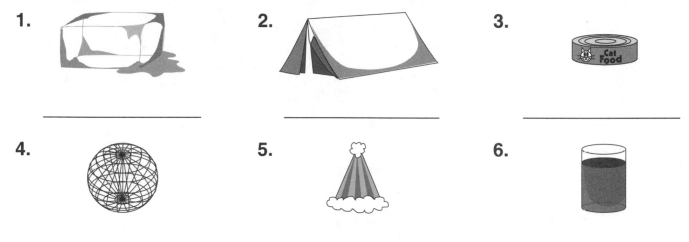

rectangular prism   cube   triangular prism   square pyramid

cylinder   cone   sphere

Write the name of the space figure each is most like.

1. _____

2. _____

3. _____

4. _____

5. _____

6. _____

Complete the table.

		name of space figure	faces	edges	vertices
7.					
8.					
9.					
10.					

PROBLEM SOLVING

11. I have 0 faces. Which space figure am I?

_____

12. I have the same number of faces and vertices. I have 8 edges. Which space figure am I?

_____

Space Figures and Polygons

A **net** is the shape made by opening a space figure and laying it flat.

This net can be folded to make a cube. It has 6 square faces.

**Use dot paper. Copy each net. Name each space figure.
Then cut, fold, and tape to make a space figure.**

1.

_____

2.

_____

Name the shape made by each cut.

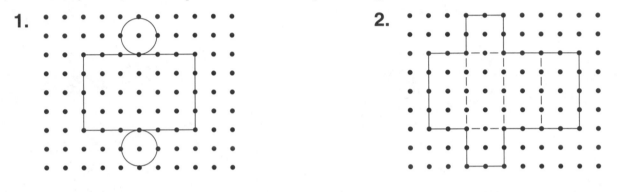

3.

_____

4.

_____

5.

_____

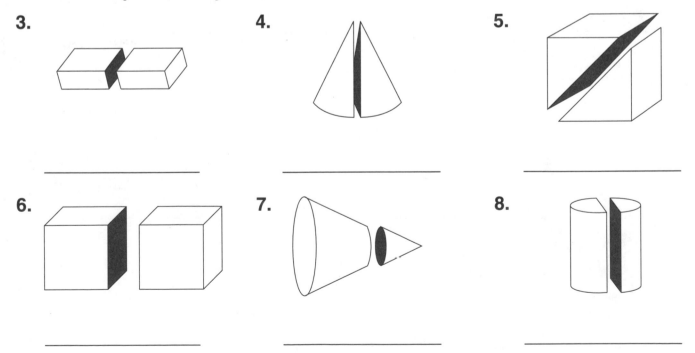

6.

_____

7.

_____

8.

_____

Volume

Name _____

Date _____

The **volume** of a space figure is the number of cubic units the figure contains. You can count cubes to find volume.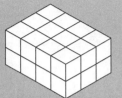

You can multiply to find volume: Volume = length × width × height

Volume =   4   ×   3   ×   2

Volume = 24 cubic units

Find the volume of each.

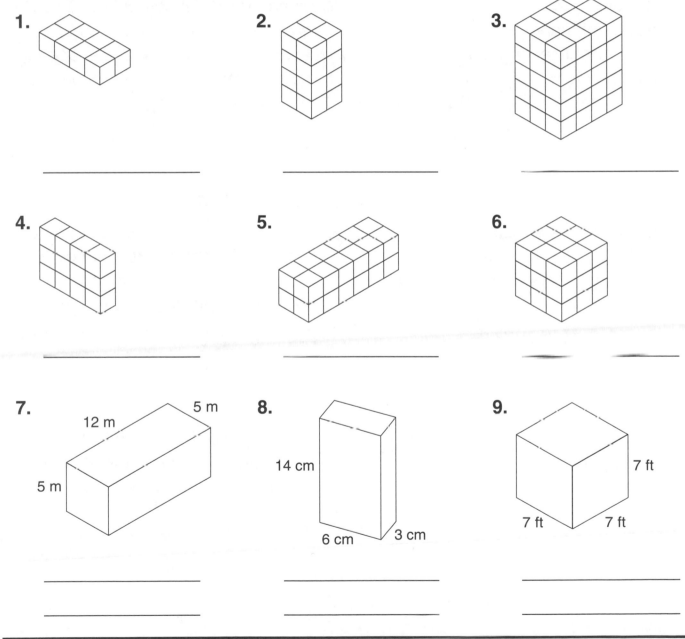

1. _____

2. _____

3. _____

4. _____

5. _____

6. _____

7. 12 m   5 m   5 m

_____

_____

8. 14 cm   6 cm   3 cm

_____

_____

9. 7 ft   7 ft   7 ft

_____

_____

Problem-Solving Strategy: Use a Drawing or Model

Name _____

Date _____

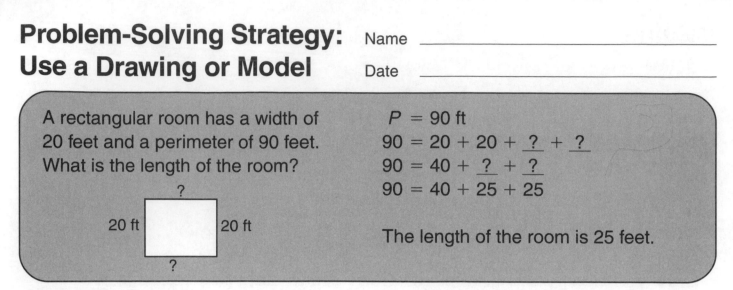

A rectangular room has a width of 20 feet and a perimeter of 90 feet. What is the length of the room?

$P = 90$ ft
$90 = 20 + 20 + \underline{?} + \underline{?}$
$90 = 40 + \underline{?} + \underline{?}$
$90 = 40 + 25 + 25$

The length of the room is 25 feet.

PROBLEM SOLVING Draw a picture or use a model.

1. A 6-pointed star is made of equilateral triangles that surround a hexagon. The length of one side of the hexagon is 4 dm. What is the perimeter of the star?

2. There are 12 fence posts around a square garden. The posts are 5 ft apart, and there is one post at each corner. What is the area of the garden? the perimeter?

_____

3. The volume of a rectangular prism is 60 cubic in. The base of the prism is 10 in. by 2 in. What is its height?

10 in.

2 in.

4. The ends of a prism are equilateral triangles that measure 7 cm on a side. The prism is 12 cm long. What is the perimeter of its net?

7 cm   12 cm
7 cm

_____

5. How many rectangles can you find in the figure below?

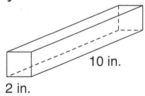

6. The length of a rectangular floor is 12 feet. Its perimeter is 42 feet. What is the area of the floor?

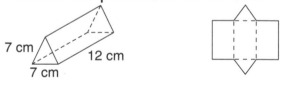

12 ft
?   ?
12 ft

_____

Division
Patterns

Fact:	$7 \div 1 = 7$	Fact:	$20 \div 5 = 4$
	$70 \div 10 = 7$		$200 \div 50 = 4$
	$700 \div 10 = 70$		$2000 \div 50 = 40$
	$7000 \div 10 = 700$		$20,000 \div 50 = 400$

Complete.

1. $5 \div 1 =$ _____

$50 \div 10 =$ _____

$500 \div 10 =$ _____

$5000 \div 10 =$ _____

2. $8 \div 2 =$ _____

$80 \div 20 =$ _____

$800 \div 20 =$ _____

$8000 \div 20 =$ _____

3. $9 \div 3 =$ _____

$90 \div 30 =$ _____

$900 \div 30 =$ _____

$9000 \div 30 =$ _____

4. $64 \div 8 =$ _____

$640 \div 80 =$ _____

$6400 \div 80 =$ _____

$64,000 \div 80 =$ _____

5. $30 \div 6 =$ _____

$300 \div 60 =$ _____

$3000 \div 60 =$ _____

$30,000 \div 60 =$ _____

6. $40 \div 5 =$ _____

$400 \div 50 =$ _____

$4000 \div 50 =$ _____

$40,000 \div 50 =$ _____

Divide mentally.

7. $420 \div 70 =$ _____

8. $4200 \div 70 =$ _____

9. $42,000 \div 70 =$ _____

10. $800 \div 40 =$ _____

11. $600 \div 20 =$ _____

12. $300 \div 10 =$ _____

13. $72,000 \div 90 =$ _____

14. $5600 \div 70 =$ _____

15. $35,000 \div 50 =$ _____

16. $5400 \div 60 =$ _____

17. $27,000 \div 90 =$ _____

18. $8100 \div 90 =$ _____

PROBLEM SOLVING

19. The dividend is 80.
The quotient is 10.
What is the divisor? _____

20. The quotient is 50.
The dividend is 2000.
What is the divisor? _____

21. The quotient is 60.
The divisor is 90.
What is the dividend? _____

22. The divisor is 50.
The quotient is 800.
What is the dividend? _____

Divisors: Multiples of Ten

$126 \div 30 = \underline{\ ?\ }$

Think: $30\overline{)126}$
Not enough ones or tens.
$30\overline{)126}$
Enough ones.

Estimate: $126 \div 30$
Try **4**.

$$\begin{array}{r} 4\ \text{R6} \\ 30\overline{)126} \\ -120 \\ \hline 6 \end{array}$$

Check.
$$\begin{array}{r} 30 \\ \times\ \ 4 \\ \hline 120 \\ +\ \ 6 \\ \hline 126 \end{array}$$

Divide.

1. $40\overline{)168}$ **2.** $20\overline{)88}$ **3.** $40\overline{)95}$ **4.** $70\overline{)296}$

5. $50\overline{)455}$ **6.** $80\overline{)342}$ **7.** $90\overline{)781}$ **8.** $60\overline{)499}$

9. $30\overline{)187}$ **10.** $40\overline{)317}$ **11.** $70\overline{)500}$ **12.** $90\overline{)627}$

13. $60\overline{)362}$ **14.** $50\overline{)298}$ **15.** $80\overline{)425}$ **16.** $30\overline{)253}$

PROBLEM SOLVING

17. The dividend is 538. The divisor
is 90. What is the quotient
and remainder?   _____

Estimating Quotients

Name _____

Date _____

Estimate: $396 \div 42 = \underline{\ ?\ }$ $400 \div 40 = 10$

396 is about 400.
42 is about 40.   So $396 \div 42$ is about 10.

Write the dividend and divisor you would use to estimate the quotient.

1. $94 \div 31$ _____

2. $78 \div 23$ _____

3. $68 \div 11$ _____

4. $\$3.70 \div 12$ _____

5. $62 \div 18$ _____

6. $91 \div 25$ _____

7. $63 \div 27$ _____

8. $211 \div 42$ _____

9. $\$5.24 \div 45$ _____

10. $965 \div 22$ _____

11. $427 \div 41$ _____

12. $630 \div 25$ _____

Estimate the quotient.

13. $86 \div 28$ _____

14. $62 \div 12$ _____

15. $\$42.05 \div 8$ _____

16. $533 \div 52$ _____

17. $\$8.63 \div 28$ _____

18. $747 \div 72$ _____

19. $4235 \div 24$ _____

20. $7381 \div 73$ _____

21. $\$23.09 \div 57$ _____

22. $\$17.31 \div 37$ _____

23. $6589 \div 71$ _____

24. $920 \div 94$ _____

PROBLEM SOLVING

25. Mike's Gas Station serviced 286 cars
in 53 days. If about the same number
of cars were serviced each day, did he
service more than 50 cars each day? _____

26. Ellen spent $6.00 to buy 32 pencils.
About how much did each pencil cost? _____

Two-Digit Dividends

Name _____

Date _____

$73 \div 22 = \underline{\ ?\ }$

Estimate:

73 ÷ **2**2

Try 3.

$$\begin{array}{r} 3\ R7 \\ 22\overline{)73} \\ -66 \\ \hline 7 \end{array}$$ ← $7 < 22$

Check.

$$\begin{array}{r} 22 \\ \times\ 3 \\ \hline 66 \\ +\ 7 \\ \hline 73 \end{array}$$

Divide and check.

1. $21\overline{)42}$ **2.** $44\overline{)88}$ **3.** $21\overline{)84}$ **4.** $23\overline{)69}$ **5.** $32\overline{)96}$

6. $54\overline{)72}$ **7.** $13\overline{)28}$ **8.** $37\overline{)41}$ **9.** $23\overline{)70}$ **10.** $46\overline{)96}$

11. $28\overline{)59}$ **12.** $33\overline{)88}$ **13.** $45\overline{)96}$ **14.** $24\overline{)75}$ **15.** $39\overline{)83}$

16. $18\overline{)\$.36}$ **17.** $27\overline{)\$.81}$ **18.** $36\overline{)\$.72}$ **19.** $24\overline{)\$.48}$ **20.** $21\overline{)\$.63}$

PROBLEM SOLVING

21. Alice divided 84 carrot sticks equally among 21 children. How many did each receive? _____

22. How many cartons can be filled with 96 eggs if each carton holds two dozen? _____

130   **Use with Lesson 12-4, text pages 388–389.**

Three-Digit Dividends

Name _____

Date _____

$169 \div 31 = \underline{\ ?\ }$

Estimate:
169 ÷ **3**1
Try 5.

$$\begin{array}{r} 5\ \text{R}14 \\ 31\overline{)169} \\ -155 \\ \hline 14 \end{array} \leftarrow \boxed{14 < 31}$$

Check.
$$\begin{array}{r} 31 \\ \times\ 5 \\ \hline 155 \\ +14 \\ \hline 169 \end{array}$$

Divide and check.

1. $62\overline{)186}$

2. $51\overline{)153}$

3. $43\overline{)172}$

4. $52\overline{)104}$

5. $22\overline{)155}$

6. $72\overline{)290}$

7. $92\overline{)369}$

8. $41\overline{)330}$

9. $32\overline{)162}$

10. $62\overline{)189}$

11. $21\overline{)149}$

12. $43\overline{)130}$

13. $81\overline{)489}$

14. $31\overline{)189}$

15. $42\overline{)129}$

16. $81\overline{)327}$

17. $62\overline{)250}$

18. $21\overline{)128}$

19. $42\overline{)252}$

20. $22\overline{)135}$

21. $53\overline{)\$2.65}$

22. $33\overline{)\$1.32}$

23. $43\overline{)\$2.58}$

24. $32\overline{)\$1.60}$

PROBLEM SOLVING

25. A cabinetmaker had 200 pieces of wood. He used 32 pieces for each cabinet. How many cabinets did he make? How many pieces of wood did he have left?

_____

Trial Quotients

$153 \div 24 =$ ___?___

Estimate:
153 ÷ **2**4
Try 7.

$$\begin{array}{r} 7 \\ 24\overline{)153} \\ 168 \end{array}$$

Too large.
Try 6.

$$\begin{array}{r} 6\text{ R}9 \\ 24\overline{)153} \\ -144 \\ \hline 9 \end{array}$$

Check.

$$\begin{array}{r} 24 \\ \times\ 6 \\ \hline 144 \\ +\ 9 \\ \hline 153 \end{array}$$

Divide.

1. $95\overline{)273}$

2. $37\overline{)222}$

3. $88\overline{)732}$

4. $26\overline{)182}$

5. $85\overline{)327}$

6. $77\overline{)448}$

7. $35\overline{)139}$

8. $59\overline{)358}$

9. $28\overline{)\$1.68}$

10. $58\overline{)\$4.06}$

11. $68\overline{)\$5.44}$

12. $48\overline{)\$2.88}$

PROBLEM SOLVING

13. A story is 450 pages long. If Pat reads 68 pages each day, how long will it take her to read the story?

_____

14. A florist had 186 roses. If he put 24 roses in each box, how many boxes did he fill? How many roses were left over?

_____

15. An album has 22 pages. Lee wants to put the same number of stamps on each page. He has 192 stamps. At most, how many can he put on each page?

_____

Use with Lesson 12-6, text pages 392–393.

Greater Quotients

$728 \div 51 = \underline{\ ?\ }$

Estimate:
728 $\div$ **51**
Try 1.

$$
\begin{array}{r}
14\ R14 \\
51\overline{)728} \\
-51\downarrow \\
\hline
218 \\
-204 \\
\hline
14
\end{array}
$$

$14 \leftarrow \boxed{14 < 51}$

Check.

$$
\begin{array}{r}
14 \\
\times\ 51 \\
\hline
14 \\
700 \\
\hline
714 \\
+\ 14 \\
\hline
728
\end{array}
$$

Divide.

1. $28\overline{)5\,8\,8}$

2. $83\overline{)9\,1\,3}$

3. $29\overline{)4\,6\,8}$

4. $36\overline{)9\,4\,9}$

5. $22\overline{)9\,1\,0}$

6. $40\overline{)9\,7\,9}$

7. $22\overline{)2\,7\,8}$

8. $32\overline{)7\,1\,2}$

9. $20\overline{)6\,3\,3}$

10. $26\overline{)9\,0\,5}$

11. $37\overline{)7\,8\,5}$

12. $62\overline{)6\,9\,9}$

13. $70\overline{)8\,1\,4}$

14. $50\overline{)6\,7\,7}$

15. $43\overline{)9\,1\,6}$

16. $46\overline{)5\,1\,3}$

17. $\$6.48 \div 12 = \underline{\quad}$

18. $\$4.20 \div 35 = \underline{\quad}$

19. $\$7.68 \div 24 = \underline{\quad}$

PROBLEM SOLVING

20. Alex spent $11.75. He bought 25 apples.
How much did each apple cost? _____

Teens as Divisors

$817 \div 13 =$ __?__

Estimate:
817 $\div$ **13**
Try 8.

$$\begin{array}{r} 8 \\ 13\overline{)817} \\ 104 \end{array}$$
Too large.
Try 7.

$$\begin{array}{r} 7 \\ 13\overline{)817} \\ 91 \end{array}$$
Too large.
Try 6.

$$\begin{array}{r} 62 \text{ R11} \\ 13\overline{)817} \\ -78\downarrow \\ 37 \\ -26 \\ 11 \end{array}$$

Check.
$$\begin{array}{r} 62 \\ \times 13 \\ \hline 186 \\ 620 \\ \hline 806 \\ +11 \\ \hline 817 \end{array}$$

Divide.

1. $11\overline{)605}$

2. $14\overline{)133}$

3. $12\overline{)226}$

4. $15\overline{)482}$

5. $11\overline{)653}$

6. $16\overline{)812}$

7. $18\overline{)987}$

8. $15\overline{)343}$

9. $13\overline{)525}$

10. $19\overline{)176}$

11. $17\overline{)189}$

12. $14\overline{)717}$

13. $16\overline{)289}$

14. $19\overline{)411}$

15. $18\overline{)381}$

16. $12\overline{)925}$

PROBLEM SOLVING

17. The divisor is 15.
The quotient is 24.
The remainder is 11.
What is the dividend? _____

18. The quotient is 35.
The divisor is 17.
The remainder is 7.
What is the dividend? _____

Four-Digit Dividends

Name _____

Date _____

$6537 \div 92 = \underline{\ ?\ }$

Estimate:
6537 $\div$ **92**
Try 7.

$$\begin{array}{r} 71\ R5 \\ 92\overline{)6537} \\ -644\downarrow \\ \hline 97 \\ -92 \\ \hline 5 \end{array}$$

Check:
$$\begin{array}{r} 71 \\ \times 92 \\ \hline 142 \\ 6390 \\ \hline 6532 \\ +5 \\ \hline 6537 \end{array}$$

Divide and check.

1. $41\overline{)1\ 5\ 8\ 1}$

2. $54\overline{)4\ 4\ 3\ 7}$

3. $75\overline{)2\ 3\ 8\ 4}$

4. $97\overline{)6\ 7\ 8\ 2}$

5. $63\overline{)5\ 3\ 5\ 7}$

6. $57\overline{)5\ 6\ 3\ 0}$

7. $43\overline{)3\ 5\ 8\ 6}$

8. $46\overline{)2\ 9\ 4\ 5}$

9. $86\overline{)1\ 6\ 4\ 6}$

10. $52\overline{)3\ 8\ 7\ 2}$

11. $77\overline{)1\ 6\ 7\ 8}$

12. $22\overline{)1\ 7\ 9\ 1}$

13. $15\overline{)1\ 2\ 4\ 3}$

14. $54\overline{)3\ 8\ 8\ 3}$

15. $82\overline{)7\ 2\ 1\ 9}$

16. $31\overline{)2\ 8\ 4\ 4}$

17. $63\overline{)5\ 3\ 2\ 1}$

18. $34\overline{)2\ 6\ 9\ 3}$

19. $61\overline{)4\ 5\ 2\ 6}$

20. $42\overline{)2\ 6\ 5\ 2}$

PROBLEM SOLVING

21. How many years are there in 1116 months? _____

22. How many days are there in 1728 hours? _____

Zero in the Quotient

$8661 \div 42 =$ ___?___

Estimate:

$8661 \div 42$

Try 2.

$$\begin{array}{r} 206 \text{ R9} \\ 42\overline{)8661} \\ -84\downarrow \\ \hline 26 \\ -\ 0\downarrow \\ \hline 261 \\ -252 \\ \hline 9 \end{array}$$

$42 > 26$
Write 0 in the tens place.

Check:

$$\begin{array}{r} 206 \\ \times\ 42 \\ \hline 412 \\ 8240 \\ \hline 8652 \\ +\ \ 9 \\ \hline 8661 \end{array}$$

Divide and check.

1. $25\overline{)1250}$ **2.** $38\overline{)760}$ **3.** $15\overline{)1050}$ **4.** $84\overline{)2563}$

5. $24\overline{)1222}$ **6.** $93\overline{)3798}$ **7.** $18\overline{)549}$ **8.** $67\overline{)4055}$

9. $46\overline{)9568}$ **10.** $27\overline{)8280}$ **11.** $21\overline{)2249}$ **12.** $63\overline{)6749}$

13. $29\overline{)2983}$ **14.** $69\overline{)8280}$ **15.** $47\overline{)9655}$ **16.** $73\overline{)7844}$

17. $52\overline{)5679}$ **18.** $39\overline{)6240}$ **19.** $92\overline{)9460}$ **20.** $49\overline{)9852}$

PROBLEM SOLVING

21. The Little League members collected 9338 labels to get new uniforms. Each of the 46 members collected the same number. How many labels did each member collect?

_____

Use with Lesson 12-10, text pages 400–401.

Problem-Solving Strategy: Hidden Information

Name _____

Date _____

Marion is baking bread for the school bake sale.
The recipe calls for $1\frac{1}{2}$ cups of milk. She has 1 pint of
milk in the refrigerator. Does she have enough milk?

Remember: 2 c = 1 pt

Compare: $2 > 1\frac{1}{2}$

Marion has enough milk.

PROBLEM SOLVING Do your work on a separate sheet of paper.

1. Samantha made 3 dozen muffins for the bake sale. She put 2 muffins in each bag. How many bags did she use?

2. Michael baked lemon squares. He tripled the recipe. Each recipe called for 1 cup of milk. How many quarts of milk did he buy?

3. Ms. Sanchez made 4 loaves of bread. She could fit 2 loaves of bread in the oven at a time. The baking time for each batch was 65 minutes. Could she bake all the loaves in 2 hours?

4. Carlos made a special dressing to eat with fresh fruit. He doubled the recipe. Each recipe called for 2 cups of yogurt. How many pints of yogurt did he buy?

5. Rashawn gave the cashier a one-dollar bill. He received a quarter, a dime, and 2 nickels as change. How much did he spend?

6. Carmela bought 8 muffins and paid with quarters. Each muffin cost 50¢. How many quarters did she give the cashier?

7. Anthony made a square sign that measured 24 inches on a side. How many feet of ribbon did he need to buy to go around the perimeter of the sign?

8. Andrea filled cups that each held 8 fluid ounces with lemonade. How many cups could she fill with each quart of lemonade?

Tenths and Hundredths

Name _____

Date _____

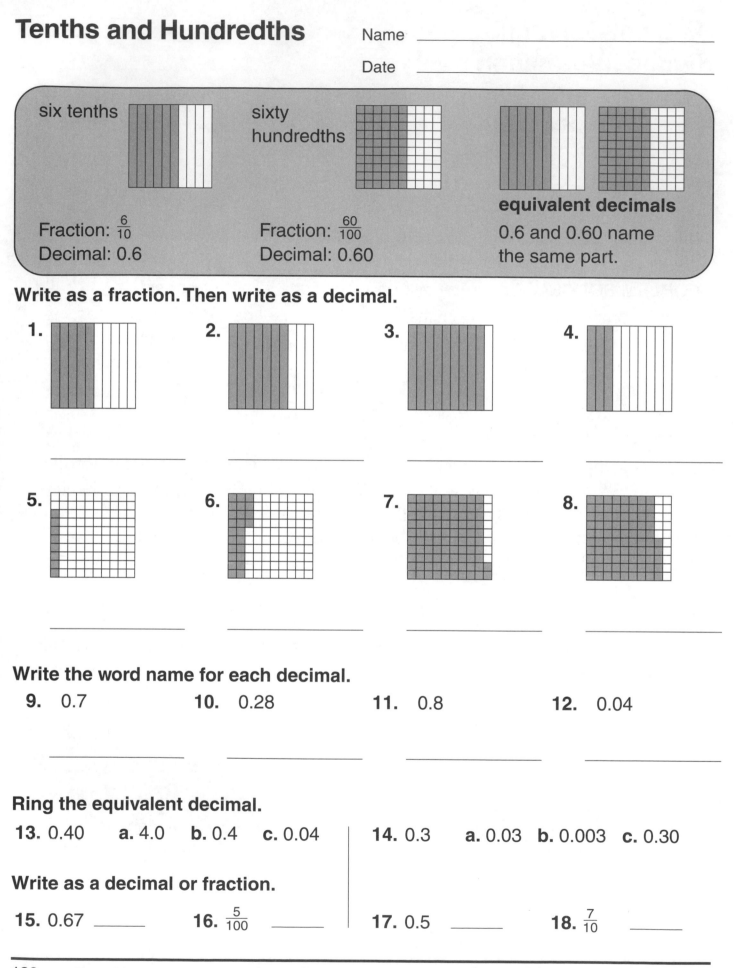

six tenths

Fraction: $\frac{6}{10}$
Decimal: 0.6

sixty hundredths

Fraction: $\frac{60}{100}$
Decimal: 0.60

equivalent decimals

0.6 and 0.60 name the same part.

Write as a fraction. Then write as a decimal.

1.

2.

3.

4.

5.

6.

7.

8.

Write the word name for each decimal.

9. 0.7   **10.** 0.28   **11.** 0.8   **12.** 0.04

Ring the equivalent decimal.

13. 0.40   **a.** 4.0   **b.** 0.4   **c.** 0.04   **14.** 0.3   **a.** 0.03   **b.** 0.003   **c.** 0.30

Write as a decimal or fraction.

15. 0.67 _____   **16.** $\frac{5}{100}$ _____   **17.** 0.5 _____   **18.** $\frac{7}{10}$ _____

Use with Lesson 13-1, text pages 412–413.

Decimals Greater Than One

Name _____

Date _____

Write as a mixed number. Then write as a decimal.

1. _____

2. _____

Write each as a decimal.

3. $54\frac{6}{100}$ _____

4. $5\frac{8}{10}$ _____

5. $22\frac{63}{100}$ _____

6. $46\frac{8}{10}$ _____

7. $3\frac{7}{10}$ _____

8. $14\frac{9}{100}$ _____

9. $71\frac{4}{100}$ _____

10. $138\frac{5}{10}$ _____

Write the word name for each decimal.

11. 5.03 _____

12. 3.98 _____

13. 48.7 _____

14. 7.9 _____

15. 4.0 _____

To which decimal is the arrow pointing?

16.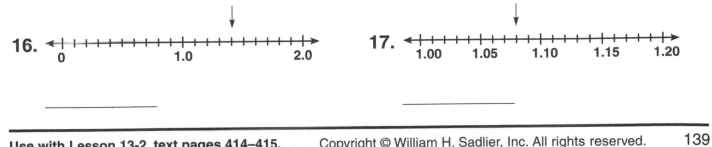

17.

_____

_____

Decimal Place Value

Name _____

Date _____

Standard Form:		Expanded Form:
32.6	⟶	30 + 2 + 0.6
8.40	⟶	8 + 0.4
837.95	⟶	800 + 30 + 7 + 0.9 + 0.05
12.09	⟶	10 + 2 + 0.09

Write the value of the underlined digit.

1. 7.84 _____ **2.** 638.15 _____ **3.** 45.61 _____

4. 15.09 _____ **5.** 163.77 _____ **6.** 31.18 _____

Write in expanded form.

7. 625.40 _____ **8.** 560.09 _____

9. 73.37 _____ **10.** 25.7 _____

11. 3.02 _____ **12.** 0.84 _____

13. 2.3 _____ **14.** 111.1 _____

Write in standard form. Then write in words.

15. 600 + 30 + 9 + 0.1 _____

16. 700 + 0.7 + 0.02 _____

17. 60 + 5 + 0.2 + 0.08 _____

18. 70 + 4 + 0.2 + 0.05 _____

19. 200 + 20 + 2 + 0.2 + 0.02 _____

20. 9 + 0.9 + 0.09 _____

Use with Lesson 13-3, text pages 416–417.

Comparing Decimals

Name _____

Date _____

Compare: 25.08 __?__ 25.6

Align the digits by place value.
Then compare the digits,
beginning at the left.

25.**08**
25.**6** $0 < 6$

So $25.08 < 25.6$

Compare. Write <, =, or >.

1. 6.29 _____ 2.69

2. 3.8 _____ 3.08

3. 7.31 _____ 7.13

4. 0.90 _____ 0.9

5. 8.4 _____ 8.36

6. 21.12 _____ 12.21

7. 406.72 _____ 406.27

8. 0.20 _____ 0.02

9. 0.88 _____ 0.87

10. 10.8 _____ 10.81

11. 328.17 _____ 328.27

12. 7.0 _____ 7

13. $0.35 _____ $3.50

14. $2.54 _____ $2.45

15. $1.03 _____ $1.30

16. 1.75 _____ 17.5

17. 4.07 _____ 4.7

18. 63.35 _____ 63.32

19. 22.1 _____ 2.21

20. 504.8 _____ 514.8

21. 8.8 _____ 8.80

22. $3.47 _____ $34.70

23. $2.63 _____ $2.67

24. $10.05 _____ $10.50

PROBLEM SOLVING

25. Ms. Asai lives in the country. She can
shop in Hayes, which is 4.25 km away
from her home. Or she can shop in
Bristol, which is 4.05 km away from
her home. Where should Ms. Asai
shop? Why?

_____

Ordering Decimals

Name _____

Date _____

Write in order from greatest to least.

1. 8.5, 8.2, 8.8 _____   **2.** 7.0, 0.07, 7.7 _____

3. 0.6, 0.9, 0.64 _____   **4.** 1.11, 11.1, 11.11 _____

5. 0.31, 3.81, 1.93 _____   **6.** 16.5, 15.6, 16.05 _____

7. 2.4, 2.39, 2.9, 2.93 _____

8. 1.57, 5.71, 1.75, 5.57 _____

9. 95.4, 95.05, 95.45, 95.5 _____

Write in order from least to greatest.

10. 5.5, 0.89, 5.39 _____   **11.** 0.9, 0.99, 0.09 _____

12. 3.1, 3.17, 3.2 _____   **13.** 6.2, 2.6, 2.2 _____

14. 1.35, 1.3, 1.03 _____   **15.** 4, 4.44, 4.4 _____

16. 0.02, 22.2, 0.2, 0.22 _____

17. 14.2, 14.02, 4.12, 4.21 _____

18. 159.5, 155.9, 159.0, 159.55 _____

Rounding Decimals

Name _____

Date _____

Round to the nearest one:		Round to the nearest tenth:	
48.73 → 49	7 > 5 Round up to 49.	48.73 → 48.7	3 < 5 Round down to 48.7.

Round to the nearest one.

1. 7.6 _____

2. 5.2 _____

3. 5.7 _____

4. 6.6 _____

5. 8.9 _____

6. 25.52 _____

7. 9.37 _____

8. 7.15 _____

Round to the nearest tenth.

9. 8.45 _____

10. 6.41 _____

11. 75.08 _____

12. 17.32 _____

13. 12.69 _____

14. 10.77 _____

15. 9.76 _____

16. 18.22 _____

17. 81.11 _____

18. 3.09 _____

19. 29.53 _____

20. 19.45 _____

Complete the table. Then answer each question.

	Winning Times of Ski Competition			
	Name	**Time in Seconds**	**Round to Nearest One**	**Round to Nearest Tenth**
21.	Maxine	46.56		
22.	Sean	44.68		
23.	Bob	47.07		
24.	Cora	41.69		

25. Who had the fastest time? _____

26. Rounded to the nearest one, whose times are the same? _____

27. Write the actual times in order from slowest to fastest. _____

Adding Decimals

Name _____

Date _____

> $13.04 + 7.9 = \underline{\ ?\ }$
>
> Line up the decimal points. Then add.
>
> Write the decimal point in the sum.
>
> $$\begin{array}{r} \overset{1}{13.04} \\ +\ 7.90 \\ \hline 20.94 \end{array}$$

Add.

1.	**2.**	**3.**	**4.**	**5.**
23.72	14.13	6	87.3	$25.04
+ 19.39	+ 5.69	+ 2.4	+ 14.5	+ 37.73

6.	**7.**	**8.**	**9.**	**10.**
7.39	32.88	37.2	64.3	$17.12
+ 3.21	+ 3.4	+ 19.6	+ 96.1	+ 84.36

11.	**12.**	**13.**	**14.**	**15.**
5.87	4.89	1.54	2.27	$7.62
+ 5.36	+ 5.56	+ 3.61	+ 9.9	+ 0.85

Align and add.

16. $20.39 + 0.72 =$ _____

17. $\$39.37 + \$8.73 =$ _____

18. $13.59 + 8.6 =$ _____

19. $\$34.30 + \$58.16 =$ _____

PROBLEM SOLVING

20. Phil jogged 1.8 km. Bill jogged 0.85 km farther than Phil. How far did Bill jog? _____

21. Katrine spent 0.75 hours painting a dresser and 1.75 hours painting her room. How much time did she spend painting? _____

22. Latisha played a video game that lasted 45.5 seconds. She played the game twice. How long did the two games last? _____

Subtracting Decimals

Name _____

Date _____

32 − 15.5 = __?__

Line up the decimal points. Then subtract.

Write the decimal point in the difference.

$$\begin{array}{r} \overset{11}{}\overset{2\ \ \cancel{3}\ 10}{3\,2.\cancel{0}} \\ -1\,5.5 \\ \hline 1\,6.5 \end{array}$$

Remember: 32 = 32.0

Subtract.

1. 31.07 − 9.30	**2.** 74.03 − 8.1	**3.** 6.45 − 5.43	**4.** 52.03 − 37.81	**5.** 60.06 − 38.2
6. 19.4 − 9.48	**7.** 96 − 69.86	**8.** 9.1 − 3.29	**9.** 57.34 − 14.54	**10.** $23.42 − 19.75
11. 8.49 − 4.54	**12.** 37.2 − 17.09	**13.** 48.23 − 4.5	**14.** $6.82 − 1.73	**15.** $7.43 − 1.95

Align and subtract.

16. 35.84 − 20.07 = _____

17. $44.55 − 32.46 = _____

18. 56.3 − 44.26 = _____

19. $12.63 − 10.97 = _____

PROBLEM SOLVING

20. The Martin family drove 32.86 km on Saturday and 19.39 km on Sunday. How much farther did the family travel on Saturday than on Sunday? _____

21. It is 89.7 km to New York and 69.9 km to New Jersey from Dan's hometown. How much farther is it to New York? _____

22. How much more is 78.4 m than 39.87 m? _____

23. Lionel ran a race in 9.34 s. Grace ran the race in 9.16 s. How much faster did Grace run the race? _____

Estimating with Decimals

Name _____

Date _____

Estimate: 63.2 + 17.9	Estimate: 40.83 − 7.3	Front-End Estimation
$\begin{array}{r} 63.2 \longrightarrow 60 \\ +\ 17.9 \longrightarrow +\ 20 \\ \hline \text{about } 80 \end{array}$	$\begin{array}{r} 40.83 \longrightarrow 40 \\ -\ 7.3 \longrightarrow -\ 7 \\ \hline \text{about } 33 \end{array}$	$\begin{array}{r} 7\,2.2\,8 \\ -\ 4\,9.7\,5 \\ \hline \text{about } 3\,0.0\,0 \end{array}$

Use rounding to estimate. Watch the signs.

1. $\begin{array}{r} 7.9 \longrightarrow \\ +\ 4.2 \longrightarrow \\ \hline \end{array}$

2. $\begin{array}{r} 24.9 \longrightarrow \\ -\ 11.8 \longrightarrow \\ \hline \end{array}$

3. $\begin{array}{r} 18.76 \longrightarrow \\ -\ 3.4 \ \longrightarrow \\ \hline \end{array}$

4. $\begin{array}{r} 6.7 \\ +\ 32.5 \\ \hline \end{array}$

5. $\begin{array}{r} 11.37 \\ -\ 3.42 \\ \hline \end{array}$

6. $\begin{array}{r} 8.6 \\ +\ 12.7 \\ \hline \end{array}$

7. $\begin{array}{r} 40.5 \\ +\ 45.34 \\ \hline \end{array}$

8. $\begin{array}{r} 76.6 \\ -\ 7.16 \\ \hline \end{array}$

9. $\begin{array}{r} 22.41 \\ +\ 7.28 \\ \hline \end{array}$

Estimate the sum or the difference. Use front-end estimation.

10. $\begin{array}{r} 15.7 \\ +\ 11.3 \\ \hline \end{array}$

11. $\begin{array}{r} 4.8 \\ -\ 3.7 \\ \hline \end{array}$

12. $\begin{array}{r} 6.1 \\ +\ 2.9 \\ \hline \end{array}$

13. $\begin{array}{r} 7.6 \\ +\ 3.4 \\ \hline \end{array}$

14. $\begin{array}{r} 0.87 \\ -\ 0.46 \\ \hline \end{array}$

15. $\begin{array}{r} 71.53 \\ -\ 36.8 \\ \hline \end{array}$

16. $\begin{array}{r} 25.31 \\ -\ 13.7 \\ \hline \end{array}$

17. $\begin{array}{r} 42.8 \\ +\ 28.2 \\ \hline \end{array}$

PROBLEM SOLVING

18. The first place relay team's time was 5.2 s.
The third place team's time was 5.58 s.
About how many seconds separate their times? _____

19. The weight of one box is 5.47 lb. Another
weighs 0.84 lb. Estimate the total weight. _____

Dividing with Money

Name _____

Date _____

What is the cost per balloon if a box of 75 costs $9?

Write a decimal point and two zeros in the dividend.	$\begin{array}{r} \$\ .12 \\ 75\overline{)\$9.0\,0} \\ -\underline{75} \\ 150 \\ -\underline{150} \\ 0 \end{array}$	Divide as usual. Write the dollar sign and decimal point in the quotient.

Each balloon costs about $.12.

Find the quotient.

1. $34 ÷ 17 = _____

2. $56 ÷ 4 = _____

3. $44 ÷ 11 = _____

4. $25 ÷ 2 = _____

5. $66 ÷ 4 = _____

6. $43 ÷ 5 = _____

7. 6)$27.

8. 5)$13.

9. 4)$10.

10. 18)$9.

11. 68)$17.

12. 16)$80.

13. 5)$19.

14. 25)$37.

PROBLEM SOLVING Tell which is the better buy.

15. 1 quart of milk for $1.00
or
4 quarts of milk for $4.20

16. 4 steaks for $19.96
or
20 steaks for $99

_____

_____

Problem-Solving Strategy: Multi-Step Problem

Name _____

Date _____

At Emerson School, 175 people ate in the cafeteria each week. During the 4 weeks in February, 26 more people ate in the cafeteria each week as well. During a special lunch, 15 more guests ate. How many people ate in the cafeteria during the month of February?

| Number of people each week: | 175 $+ 26$ 201 | Number of people for 4 weeks: | 201 $\times 4$ 804 | Total: | 804 $+ 15$ 819 |

During the month of February, 819 people ate in the cafeteria.

PROBLEM SOLVING Do your work on a separate sheet of paper.

1. An auditorium had 20 rows of seats, with 18 seats in each row. Only the first 6 rows were filled with people. How many seats were empty?

2. Eighty people want to contribute towards a $100 gift. If each person gives $2, how much change did each receive?

3. Ms. Fernandez bought lunch for 6 children. Each child had a salad, a sandwich, and a drink. Each salad cost $1.95, and each sandwich cost $4.75. She spent $7.25 on drinks for the group. What was the total cost of the childrens' lunch?

4. Ten friends decided to go on a camping trip. They bought 3 tents that cost $80 each. Each person spent $30 on food. A camping permit for the group cost $60. If the expenses were shared equally, how much did each person spend?

5. Five members of a team each drove to a swim meet. Each person spent $10 on gas. Tolls for each car were $1.25. How much money could they have saved if they all went in one car?

6. Robert's board was 135.5 cm long. He cut off 2 pieces that each measured 50.5 cm. He cut off another piece that was 25.5 cm. After he cut the board, how long was it?

Problem Solving: Review of Strategies

Name _____

Date _____

Solve. Do your work on a separate sheet of paper.

1. 6 is to $\frac{1}{2}$ and $\frac{1}{3}$ as 12 is to ____ and ____.

2. 212°F is to 32°F as 100°C is to ____.

3. Ralph has 2 cats, Stanley and Jenny. Together they weigh $23\frac{7}{8}$ lb. Jenny weighs $3\frac{7}{8}$ lb more than Stanley. How much does Stanley weigh?

4. Andrea's rectangular garden has an area of 160 square feet. One side of the garden is 8 feet long. What is the perimeter of the garden?

5. Pedro is making a collage using a quadrilateral, a triangle, and an octagon. The first shape he pastes onto the collage has fewer sides than the second shape, but more sides than the third. In what order does he paste the shapes?

6. Mr Mayer wants to buy a dozen baseball caps for $4.95 each and half a dozen bats for $29.95 each. He has a budget of $225. How much does the equipment cost? Can he buy the equipment and stay within the budget?

7. Tony draws a pentagon. Each side is 40 cm long. Is its perimeter more or less than a meter? Explain.

8. Mr. DeRoche bought 5 pens for $2.29 each and 3 notebooks for $2.59 each. He paid with a twenty-dollar bill. How much change did he receive?

9. $\frac{1}{2}$ is to $\frac{6}{12}$ as $\frac{1}{3}$ is to $\frac{?}{12}$.

10. 72 is to 8 as 45 is to ____.

Number Sentences

Name _____

Date _____

Mr. Foster travels 2 days at 300 miles a day to get to the island.
How many miles does he travel?

Write a number sentence to help you solve the problem.
• Let n stand for the number you want to find.
• Write the number sentence. $2 \times 300 = n$
• Solve for n. $600 = n$
                                      He traveled 600 miles to the island.

PROBLEM SOLVING Write a number sentence. Then solve.

1. There were two mountains on the island. One was 1760 feet tall, the other 3839 feet tall. What was the difference in height of the two mountains?

_____

2. Rangers allow 240 cars to drive up a mountain each hour. If this number of cars drove up every hour between 12 noon and 4:00 P.M., how many cars made the climb?

_____

3. The island is on a lake that is 1500 feet above sea level. How far above sea level is the top of each mountain? (See Problem 1.)

_____

4. There are four walking trails up the mountain. They are 0.9 mi, 1.3 mi, 1.9 mi, and 2.1 mi long. What is the total length of these trails?

_____

5. One ranger offers a cliff climbing tour that takes one hour. He takes 6 people on each tour. How many tours will he give for 72 people?

_____

6. A full day of activities at the mountain costs $14.95 per person. What did a group of 8 people pay for one day of activities?

_____

7. A group of 10 people received a discount of $2.00 per person off the regular amount of $14.95. What was the total cost for the group?

_____

8. The gift shop had 492 key chains. Each rack could hold 75 key chains. What is the least number of racks needed for all the key chains?

_____

Finding Missing Numbers

Name _____

Date _____

What number does y stand for? $24 - 10 = 7 \times y$

Compute.	**Solve.**	**Check.**
$24 - 10 = 7 \times y$	$14 \div 7 = y$	$24 - 10 = 7 \times 2$
$14 = 7 \times y$	$14 \div 7 = 2$	$\quad 14 \quad = \quad 14$
	$y = 2$	

Write the number that y, n, or x stands for in each number sentence.

1. $x - 3 = 5 + 5$ 　　　　**2.** $6 \times 5 = 17 + y$ 　　　　**3.** $47 - n = 10 \times 3$

_____ 　　　　_____ 　　　　_____

4. $100 \div 2 = x + 27$ 　　　**5.** $632 - 1 = y + 631$ 　　　**6.** $x - 20 = 72 - 12$

_____ 　　　　_____ 　　　　_____

7. $20 \times 2 + 14 = 9 \times n$ 　　　　　　**8.** $160 - 80 - 4 = y \times 2$

_____ 　　　　　　　　_____

9. $62 - 50 - y = 4 \times 3 - 12$ 　　　**10.** $10 \times 10 \times 10 = 3 \times n + 700$

_____ 　　　　　　　　_____

11. $100 \times y + 5 = 20 \times 5 + 5$ 　　　**12.** $x + 47 - 2 = 7 \times 7 - 1$

_____ 　　　　　　　　_____

Functions

Name _____

Date _____

Input	Rule		Output		Input	Rule		Output	
16	÷ 4	=	4		22	+ 11	=	33	
12	÷ 4	=	3		33	+ 11	=	44	

Complete each function table.

1.

Rule: × 6	
Input	Output
7	
9	
11	
13	
15	

2.

Rule: + 90	
Input	Output
	93
	103
	123
	153
	193

3.

Rule: − 5	
Input	Output
68	
73	
78	
83	
88	

4.

Rule: ÷ 8	
Input	Output
96	
	11
72	
	8
56	

5.

Rule: × 22	
Input	Output
3	
	110
7	
10	
	286

6.

Rule: − 13	
Input	Output
	27
60	
75	
	72
90	

Write the rule for each table.

7.

Rule: ?	
Input	Output
161	23
175	25
189	27
203	29
217	31

8.

Rule: ?	
Input	Output
18	37
36	55
54	73
72	91
90	109

9.

Rule: ?	
Input	Output
21	1050
24	1200
30	1500
42	2100
66	3300

_____   _____   _____

Use with Lesson 14-3, text pages 446–447.

Missing Symbols

Name _____

Date _____

Which symbol completes the number sentence, = or ≠ ?

Compute both sides.

5×6 _?_ $40 - 10$

30 _?_ 30

Compare. $30 = 30$

Compute both sides.

$80 \div 8$ _?_ $8 + 3$

10 _?_ 11

Compare. $10 \neq 11$

Compare. Write = or ≠.

1. $16 \div 2$ _____ $100 - 92$

2. 4×30 _____ $30 + 40 + 50$

3. $\$7.50 + \$16.00$ _____ $\$30.00 - \$4.50$

4. $\$12.33 \div 3$ _____ $\$28.77 \div 7$

5. 7×25 _____ 10×17

6. $735 \div 5$ _____ $110 + 37$

7. $543 - 345$ _____ $1845 - 1647$

8. 7×66 _____ 6×77

9. $1600 \div 40$ _____ $1500 - 30$

10. $3986 + 8738$ _____ 123×103

Compare. Write <, =, or >.

11. $56 + 78$ _____ $124 - 32$

12. 43×27 _____ $565 \div 5$

13. $400 \div 20$ _____ 20×1

14. $24 - 3$ _____ $2 + 20$

15. $2456 + 301$ _____ 90×30

16. 3×700 _____ $4200 \div 2$

17. 40×60 _____ 50×50

18. $353 + 535$ _____ $1776 \div 2$

Number Line

Let $n = 19$. What is $n + 34$?

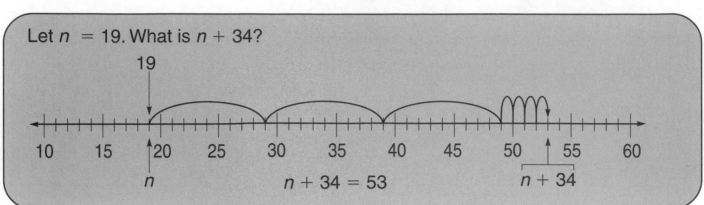

n $n + 34 = 53$ $n + 34$

Complete. You may use number lines.

1. $n = 50$
$n + 25 =$ _____

2. $n = 207$
$n - 86 =$ _____

3. $n = 643$
$n - 296 =$ _____

4. $n = 536$
$n + 88 =$ _____

5. $n = 99$
$n + 364 =$ _____

6. $n = 125$
$n - 50 =$ _____

7. $n = 86$
$n - 49 =$ _____

8. $n = 53$
$n + 58 =$ _____

9. $n = 100$
$n + 45 =$ _____

10. $n = 100$
$n - 56 =$ _____

Complete each table. You may use number lines.

11.

n	$n + 19$
17	
23	
39	
57	

12.

n	$n - 26$
42	
51	
73	
93	

13.

n	$n + 47$
111	
100	
89	
63	

Using Parentheses

Name _____

Date _____

Evaluate:

$$(16 - 10) \times 8 \div 2 - 4 + 30 = \underline{?}$$
$$6 \quad \times 8 \div 2 - 4 + 30 = \underline{?}$$
$$48 \quad \div 2 - 4 + 30 = \underline{?}$$
$$24 \quad - 4 + 30 = \underline{?}$$
$$20 \quad + 30 = 50$$

- Do operations in parentheses first.
- Multiply or divide.
  Work in order from left to right.
- Add or subtract.
  Work in order from left to right.

Use the order of operations to solve.

1. $(5 \times 3) + (7 - 2)$

2. $(12 \div 3) + (16 \div 4)$

3. $(81 \div 9) \div (56 - 53)$

4. $(3 \times 34) - (4 \times 17)$

5. $(3 \times \$2.25) + (\$10.00 \div 4)$

6. $(3.9 + 4.0) - 5 + 16$

7. $3 \times (9 - 1) - 12 - (4 \times 3)$

8. $(100 \div 4) \times 4 \div 4$

9. $\left(\frac{6}{10} - \frac{1}{2}\right) + \left(\frac{3}{10} + \frac{1}{2}\right)$

10. $\left(\frac{3}{7} + \frac{2}{7}\right) \times \left(\frac{4}{7} + \frac{3}{7}\right)$

Problem-Solving Strategy: More Than One Way

Name _____

Date _____

Peter needs a 64 sq. ft. platform for his electric train. He has two sheets of plywood that are each 4 ft by 8 ft. If he puts them together will the area be big enough?

Draw a picture.

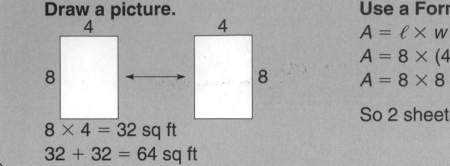

$8 \times 4 = 32$ sq ft

$32 + 32 = 64$ sq ft

Use a Formula.

$A = \ell \times w$

$A = 8 \times (4 + 4)$

$A = 8 \times 8 = 64$ sq ft

So 2 sheets are enough.

Solve. Do your work on a separate sheet of paper.

1. Peter is setting up an oval train track 56 feet long. Each section of track measures $\frac{1}{2}$ foot. How many sections does he need?

2. Peter places three road signs at each of 12 crossings and 6 road signs at each of the 5 crossings. How many road signs are there?

3. There are 16 areas where Peter wants miniature trees. If he has 243 trees, can he put 15 trees in each area?

4. Peter divided his train cars into 3 groups. The first group had 15 cars. The groups had the same number of cars. How many train cars did he have?

5. Peter has 6 ft of flexible fence. Does he have enough to put around a model playground that is a 10-in. by 16-in. rectangle?

6. Peter has twice as many flatcars as coal cars and 3 times as many boxcars as flatcars and coal cars together. He has 8 coal cars. How many boxcars does Peter have?